This book is the first systematic and rigorous account of continuum percolation. The authors treat two models, the Boolean model and the random connection model, in detail, and discuss a number of related continuum models. Where appropriate, they make clear the connections between discrete percolation and continuum percolation.

All important techniques and methods are explained and applied to obtain results on the existence of phase transitions, equality of certain densities, continuity of critical densities with respect to distributions, uniqueness of the unbounded component, covered volume fractions, compression, rarefaction and so on. The book is self-contained, assuming only familiarity with measure theory and basic probability theory. The approach makes use of simple ergodic theory, but the underlying geometric ideas are always made clear.

CAMBRIDGE TRACTS IN MATHEMATICS

General Editors
B. BOLLOBAS, P. SARNAK, C. T. C. WALL

119 Continuum percolation

RONALD MEESTER
University of Utrecht

RAHUL ROY
Indian Statistical Institute

Continuum percolation

CAMBRIDGE UNIVERSITY PRESS
Cambridge, New York, Melbourne, Madrid, Cape Town, Singapore, São Paulo

Cambridge University Press
The Edinburgh Building, Cambridge CB2 8RU, UK

Published in the United States of America by Cambridge University Press, New York

www.cambridge.org
Information on this title: www.cambridge.org/9780521475044

© Cambridge University Press 1996

First published 1996
This digitally printed version 2008

A catalogue record for this publication is available from the British Library

Library of Congress Cataloguing in Publication data
Meester, Ronald.
Continuum percolation / Ronald Meester, Rahul Roy.
p.   cm.
Includes bibliographical references and index.
ISBN 0-521-47504-X (hc)
1. Percolation (Statistical physics)   2. Stochastic processes.
I. Roy, Rahul.   II. Title.
QC174.85.P45M44   1996
530.1'3 – dc20   95–40924
                                              CIP

ISBN 978-0-521-47504-4 hardback
ISBN 978-0-521-06250-3 paperback

Contents

Preface   *page* ix

1   Introduction   1
1.1   Motivation for continuum models   1
1.2   Discrete percolation   2
1.3   Stationary point processes   9
1.4   The Boolean model   14
1.5   The random-connection model   18
1.6   Notes   20

2   Basic methods   21
2.1   Ergodicity   21
2.2   Coupling and scaling   28
2.3   The FKG inequality   31
2.4   The BK inequality   34
2.5   Notes   39

3   Occupancy in Poisson Boolean models   40
3.1   Introduction   40
3.2   One-dimensional triviality   43
3.3   Critical phenomena   45
3.4   Critical densities   52
3.5   Equality of the critical densities   59
3.6   Uniqueness   63
3.7   Exponential decay   68
3.8   Continuity of the critical density and the percolation function   71

3.9   Bounds on λ_c and asymptotics for the cluster size   85
3.10   Notes   90

4   Vacancy in Poisson Boolean models   91
4.1   Critical densities   92
4.2   RSW – notation and definition   95
4.3   RSW – construction   97
4.4   RSW – preliminary results   103
4.5   RSW – proof   106
4.6   Equality of the critical densities   108
4.7   Uniqueness   116
4.8   Continuity of the percolation function   119
4.9   Notes   121

5   Distinguishing features of the Poisson Boolean model   122
5.1   The covered volume fraction   122
5.2   Compression   125
5.3   Rarefaction   138
5.4   Notes   150

6   The Poisson random-connection model   151
6.1   Non-triviality of the model   152
6.2   Properties of the connection function   155
6.3   Equality of the critical densities   159
6.4   Uniqueness   172
6.5   High density   173
6.6   Notes   180

7   Models driven by general processes   181
7.1   Ergodic decomposition   181
7.2   Basic facts on coverage   184
7.3   Unbounded components in Boolean models   187
7.4   Uniqueness in Boolean models   194
7.5   Uniqueness in random-connection models   197
7.6   Cutting and stacking   199
7.7   Examples   204
7.8   Notes   208

8	**Other continuum percolation models**	209
8.1	Continuum fractal percolation	209
8.2	Percolation of level sets in random fields	216
8.3	Dependent Boolean and random-connection models	221
8.4	Stationary spanning forests	226
8.5	Percolation of Poisson sticks	229
8.6	Notes	232
References		233
Index		237

Preface

This is the first book completely devoted to continuum percolation. The idea to write this book came up after we noticed that even specialists working in the larger area of spatial random processes were unaware about the current state of the art of continuum percolation. Although stochastic geometers have extensively studied the Boolean model, which is one of the most common models of continuum percolation, their focus has been on geometric and statistical aspects rather than on percolation-theoretical issues.

Initially, we planned to write a review article, but it became clear very quickly that it would be impossible to cover even the most basic results in such a review. Also it became apparent that it would be impossible to include in one volume all available results of a subject this size and still expanding. Therefore, we decided on a book which would give attention to all major issues and techniques without necessarily pushing them to the frontier of today's knowledge. When there is more to say on a specific subject than is found here, we provide the appropriate references for further reading.

Continuum percolation models are easily described verbally, but unlike discrete percolation models, their formal mathematical construction is not completely straightforward. In fact, many people (the authors included) have been quite careless with these constructions in the literature. The setup we have chosen in this book is probably the simplest rigorous construction which allows us to use all the ergodic theory we want. Perhaps some people will be bothered by the fact that we define ergodicity in terms of discrete group actions rather than as a continuum, but we prefer to avoid measure-theoretical nightmares in a book which is supposed to be on percolation theory and in which ergodicity is just a tool to work with. This rigorous construction is more for the mathematical completeness of the book. The reader may easily understand the book with just the geometric notion of the models in mind.

If we compare the results in this book to well-known results in discrete percolation models, we can roughly distinguish three classes of results. The first class of results consists of those for which there is a natural analogue in discrete percolation. For this class of results, the reader will notice that there are usually extra technical complications because we work in the continuum. These complications are often topological in nature or have something to do with the dependency structure in the models. As a result, statements are usually not quite the same as their discrete counterpart. Examples in this class are the non-triviality of phase transitions, the RSW lemma, the uniqueness of unbounded components and the equality of certain critical densities in random connection models. The second class of results consists of those whose discrete counterpart is either false or unknown. Examples in this class are the fact that the two most natural critical densities in certain Boolean models are not equal, the possible non-uniqueness of the unbounded component in certain cases and some limit results in continuum fractal percolation. And then, of course, there is a class of results which do not have a discrete analogue at all and which are special features of continuum models. This class, fortunately, is quite large and contains all high-density results like compression and rarefaction, results on the covered volume fraction, continuity of the critical densities when the radii distributions converge weakly in Boolean models, scaling properties and complete coverage results. In discrete models, the 'dual' structure plays an important role and can be described independently of the original structure. In continuum models, the vacancy structure plays a role similar to that of the dual, but the vacancy structure can only be described through the occupancy structure as its complement.

We have tried to make the book as self-contained as possible. It helps if the reader is familiar with discrete percolation theory, but this is not really necessary. The reader should feel comfortable with measure theory and basic probability theory, including branching processes. In order to avoid references scattered throughout the text, we conclude each chapter with notes which contain background information and references to related material.

Finally, we would like to thank Olle Häggström, Remco van der Hofstad, Karin Nelander, Mathew Penrose, Anish Sarkar and Jeffrey Steif for many comments on drafts of this book varying from correcting language to pointing out serious mathematical mistakes. Several visits of Rahul Roy to The Netherlands were partly financed by The Dutch Organisation of Scientific Research (N.W.O.). Ronald Meester would like to thank the people in the Indian Statistical Institute in New Delhi for their hospitality during several visits.

October 1995   Ronald Meester
                                                         Rahul Roy

1

Introduction

1.1 Motivation for continuum models

Many phenomena in physics, chemistry and biology can be modelled by spatial random processes where the randomness is in the geometry of the space rather than in the random behaviour or motion of an object in a deterministic setting. As typical examples of the phenomena we have in mind, consider the spread of a disease in an orchard where the trees are arranged in a grid, and where the disease spreads from an infected tree to its neighbouring trees. In this example, the owner of the orchard is interested in the probability that a particular disease will eventually kill all the trees in the orchard. Another example is the process of the ground getting wet during a period of rain. The randomness here is the place where the raindrops fall on the ground and the size of the wetted region per raindrop. Finally, consider the spread of a disease in a forest. The infection is transmitted from one tree to another, which need not be in the vicinity of the infected tree. This is more likely to happen when the trees are closely spaced than when they are far apart. The collection of infected trees forms a random subset of trees in the forest.

The geometric structure of the first example is discrete, whereas in the next two examples, although the number of raindrops or trees is countable, the position of either is in the continuous space. A rigorous mathematical model to describe the first example is the standard discrete percolation model. This model has been studied extensively in the last three decades and an excellent reference on the mathematical aspects of this model is the book by Grimmett (1989). The second and third examples are usually described by a continuum percolation model; such models are the subject of this book.

Some geometric aspects of the continuum percolation model have been studied in the context of stochastic geometry. In the language of stochastic geometry,

1

the continuum percolation model is usually referred to as a coverage process or a Boolean model.

To get an idea of the kind of questions addressed in a percolation-theoretic study, we elaborate the examples of rainfall and the spread of a disease in a forest. Before the rain starts, the ground is assumed to be completely dry. At a point where a raindrop falls, the ground soaks up the water and a circular wet patch is formed. When the first raindrops fall, we see small wet regions inside a large dry region. The wet region grows when more raindrops reach the ground and at some instant, so many raindrops have reached the ground that the picture suddenly changes from wet 'islands' inside a large dry region to dry 'islands' inside a large wet region. This phenomenon of a sudden drastic change in the global spatial structure is called a *phase transition*. Typically, the parameter of the model is not the time, but the density of raindrops on the ground. So for instance, we say that the phase transition takes place at a given density of the raindrops, rather than at a given time. The nature of such phase transitions is an important subject in the percolation-theoretic study of Boolean models. In the example of the spread of an infection in a forest, a question of interest is whether the infection of one particular tree may result in the infection being transmitted to a tree far away. This is of course more likely when the density of the trees in the forest is high. Based on the density of the trees, a phase transition formulation may be obtained for this model too.

The focus of this book is on mathematically rigorous results in models of continuum percolation. In this context, we remark that there are many results available in the applied literature which have yet to be mathematically verified.

1.2 Discrete percolation

Before we introduce continuum percolation models we present a short treatment of discrete models. There are several reasons for doing this. First, independent percolation on the integer lattice (to be defined below) was the first percolation model studied and many of the ideas in the theory of this model can be used in the study of continuum models as well; many of the results in the continuum are analogues of discrete results. Secondly, discrete percolation models are in some sense the simplest percolation models to describe and they are suitable for the reader to get a feeling for the types of problems which are involved. Finally, an important technique in the theory of continuum percolation is to approximate the continuum model by a discrete one. In these instances, we need to know something about discrete percolation. Our treatment of discrete percolation is concise, and we refer the reader to Grimmett (1989) for a detailed discussion of discrete percolation. For proofs which are not given here we refer to either Grimmett (1989) or the references in the notes.

The setup is as follows. Each element of the d-dimensional integer lattice $\mathbf{Z}^d$ is a vertex, where $d \geq 1$. Two vertices at a Euclidean distance one apart are called *neighbours*. Each pair of neighbours has an edge between them. The graph obtained this way is, with a slight abuse of notation, denoted by $\mathbf{Z}^d$. An edge is often called a *bond*, and the set of all bonds is denoted by $\mathbf{E}$. Bonds can be either *open* or *closed*. A *path* is a finite or infinite alternating sequence $(z_1, e_1, z_2, e_2, \ldots)$ of vertices z_i and bonds e_i such that $z_i \neq z_j$ and $e_i \neq e_j$ whenever $i \neq j$ and such that e_i is the bond between the neighbours z_i and z_{i+1}, for all i. The length of a path is the number of bonds it contains. A *circuit* is a finite path, the only difference being that it starts and ends at the same vertex. An *open (closed) path* is a path whose bonds are all open (closed). Two vertices are said to be *connected* if there is a finite open path from one to the other. An open *cluster* is a set of connected vertices which is maximal with respect to this property. Of course, clusters can be either finite or infinite. The open cluster containing the origin is denoted by $C(0)$.

Next we introduce probability. For $0 \leq p \leq 1$, we equip the space $\Omega = \{0, 1\}^{\mathbf{E}}$ with the natural product measure P_p, which is defined via $P_p(\omega(e) = 1) = p$ for all $e \in \mathbf{E}$. For any realisation $\omega \in \Omega$, the bond e is said to be open if $\omega(e) = 1$ and closed otherwise. Thus each bond is open with probability p independently of all other bonds.

This is the basic percolation model on the d-dimensional integer lattice. We are interested in unbounded clusters, so here are some natural definitions:

Definition 1.1 *The percolation function $\theta^{(d)}$ is defined by*

$$\theta^{(d)}(p) = P_p(\mathrm{card}(C(0)) = \infty).$$

We define the function $\chi^{(d)}(p)$ by

$$\chi^{(d)}(p) = E_p(\mathrm{card}(C(0))),$$

where E_p is the expectation operator corresponding to P_p, and $\mathrm{card}(\cdot)$ denotes cardinality.

Much of the theory of discrete percolation is concerned with the behaviour of these functions. It seems obvious that both $\theta^{(d)}$ and $\chi^{(d)}$ are non-decreasing in p. In Chapter 2 it will become clear how to prove this. Based on $\theta^{(d)}$ and $\chi^{(d)}$ we can define the following *critical probabilities*:

Definition 1.2 *The critical probability $p_c(d)$ is defined by*

$$p_c(d) = \inf\{p : \theta^{(d)}(p) > 0\}.$$

The critical probability $p_T(d)$ is defined by

$$p_T(d) = \inf\{p : \chi^{(d)}(p) = \infty\}.$$

The two critical probabilities just defined are quite natural. There is a third critical probability which may not seem that natural at first sight but which turns out to be very useful. To define this critical probability, let $\sigma_p((n_1, n_2, \ldots, n_d), i)$ be the probability that the box $[0, n_1] \times [0, n_2] \times \cdots \times [0, n_d]$ contains an open path connecting two opposite faces in the i-th direction.

Definition 1.3 *For $d \geq 2$, the critical probability $p_S(d)$ is defined by*

$$p_S(d) = \inf\{p : \limsup_{n \to \infty} \sigma_p((n, 3n, 3n, \ldots, 3n), 1) = 0\}.$$

It is obvious that $p_T(d) \leq p_c(d)$ for all d. It is also clear that $p_c(1) = p_T(1) = 1$. Other properties of these critical probabilities are not so easy to obtain:

Theorem 1.1 *For all $d \geq 2$ we have $0 < p_c(d) < 1$.*

Theorem 1.2 *For all $d \geq 2$ we have $p_c(d) = p_T(d) = p_S(d)$.*

Actual values are known only in one and two dimensions. It is obvious that $p_c(1) = 1$ and we also know that $p_c(2) = \frac{1}{2}$. This last result is far from trivial! The proof of Theorem 1.2 is very hard and we do not give it here. Theorem 1.1 lies at the very heart of percolation theory. It establishes the existence of a *phase transition*; i.e. the macroscopic behaviour of the system is very different for values of p below and above the critical probability $p_c(d)$. If an infinite cluster exists we say that *percolation* occurs. The idea behind the proof of Theorem 1.1 will be used a few more times in this book, so we present the proof here:

Proof of Theorem 1.1 The inequality $p_c(d) > 0$ is very simple. Indeed, the number of distinct paths of length n starting at the origin is at most $2d(2d - 1)^{n-1}$. (For the first bond, we have $2d$ possibilities; after that we have at most $2d - 1$ possibilities for each new bond because we are not allowed to go back to where we came from.) Each of these paths has probability p^n to be open. Thus the expected number of open paths of length n starting at the origin is at most $2d(2d - 1)^{n-1} p^n$. If $p < (2d - 1)^{-1}$ then $\sum_{n=1}^{\infty} 2d(2d - 1)^{n-1} p^n < \infty$, and hence the expected number of open edges in the component $C(0)$ is finite. This necessarily means that the probability that the component $C(0)$ is finite is equal to 1 and hence $\theta^{(d)}(p) = 0$ if $p < (2d - 1)^{-1}$. Thus we obtain that $p_c(d) \geq (2d - 1)^{-1}$.

For the other inequality we observe that it suffices to prove it for the case $d = 2$ as $p_c(d)$ is clearly non-increasing in d. We need to introduce the *dual* graph $\mathbf{Z}^{2*}$. This is the graph obtained from $\mathbf{Z}^2$ by shifting it by the vector $(\frac{1}{2}, \frac{1}{2})$. The set of edges of the dual graph is denoted by $\mathbf{E}^*$. Each edge of $\mathbf{E}$ now crosses exactly one edge in $\mathbf{E}^*$. We declare an edge in $\mathbf{E}^*$ to be open if and only if the edge it crosses in $\mathbf{E}$ is open, and closed otherwise. It is intuitively obvious and a well-known fact in graph theory (Whitney 1933) that there is a closed circuit in $\mathbf{Z}^{2*}$ surrounding the origin if and only if $C(0)$ is finite. Now we can perform a counting argument as in the first part of this proof. There are at most $n3^n$ distinct circuits of length n surrounding the origin. (This is a rather crude bound: such a circuit has to contain at least one vertex on the x-axis. There are at most n possibilities for this. Starting at this vertex, we have only three possibilities for each new bond.) If for some $N > 0$ (i) all bonds in $[-N, N] \times [-N, N]$ are open and (ii) there is no closed circuit in the dual surrounding $[-N, N] \times [-N, N]$, then $C(0)$ is infinite. The event in (i) certainly has positive probability. Furthermore, a circuit surrounding $[-N, N] \times [-N, N]$ has length at least $4N$. Hence, if $p > \frac{2}{3}$ we can choose N so large that $\sum_{n=4N}^{\infty} n3^n(1 - p)^n < 1$ and for such p and N, the event in (ii) also has positive probability. Because the events in (i) and (ii) depend on disjoint sets of edges, they are independent, and we conclude that $p_c(2) \leq \frac{2}{3}$. $\qquad\square$

Now that we have established the existence of infinite open clusters for $p > p_c$, the question arises of just how many infinite open clusters exist. There is a remarkable answer to that question. First observe that the existence of an infinite open cluster does not depend on the state of any finite set of bonds. Hence it follows from Kolmogorov's $0 - 1$ law that the existence of an infinite open cluster has probability either zero or one. This, of course, corresponds to the different phases of the percolation model: for $p < p_c(d)$ there is no infinite open cluster a.s. and for $p > p_c(d)$, the probability of having an infinite open cluster is positive and hence equal to 1. What happens *at* the critical probability is known in two dimensions and in dimension higher than 19 only: there is no infinite cluster a.s. in these cases. The remarkable fact referred to above is the following:

Theorem 1.3 *There is at most one infinite open cluster a.s.*

This result is referred to as the *uniqueness of the infinite cluster*.

We continue the discussion with some basic inequalities which are very useful in the analysis of models of this type. For this, we need to introduce some terminology. There is a natural partial order on $\Omega = \{0, 1\}^{\mathbf{E}}$: $\omega \le \omega'$ if and only if $\omega(e) \le \omega'(e)$ for all $e \in \mathbf{E}$. An event A in Ω (we assume that Ω is equipped with the usual Borel σ-field) is said to be *increasing* if its indicator function is increasing (a real-valued function f on Ω is increasing if $f(\omega) \le f(\omega')$ whenever $\omega \le \omega'$). An event A is said to be *decreasing* if its complement is increasing. A typical example of an increasing event is the event that two distinct vertices are connected to each other by an open path.

More generally, we can consider a product space $\Omega_k = \{0, 1, \ldots, k\}^{\Sigma}$, where Σ is a countable set, and equip Ω_k with product measure $P_p = \{p_0, p_1, \ldots, p_k\}^{\Sigma}$, where $\sum_{i=0}^{k} p_i = 1$. There is a natural partial order on Ω_k and the notions of increasing and decreasing events generalise easily. Writing P_p and E_p for probabilities and expectations with respect to p, we have the following important inequality:

Theorem 1.4 (FKG inequality) *Let f_1 and f_2 be both increasing or both decreasing functions. Then*

$$E_p f_1 f_2 \ge E_p f_1 E_p f_2.$$

Taking f_1 and f_2 to be the indicator functions of two increasing (or two decreasing) events A and B, respectively, this inequality reduces to

$$P_p(A \cap B) \ge P_p(A) P_p(B).$$

This result is not surprising: if there exists an open path connecting two different vertices, another path connecting two other vertices becomes more likely as it can 'use' the bonds of the first path.

Sometimes we need an inequality which goes in the opposite direction. Given the existence of an open path connecting two vertices, we can make it 'harder' for other connections to exist by requiring them to be disjoint from the first connection. This motivates the following definition. Suppose A and B are increasing events which depend only on the state of finitely many bonds. We define $A \square B$ to be the set of all configurations ω for which there exist disjoint sets of open bonds with the property that the first such set guarantees the occurrence of A and the second guarantees the occurrence of B. More precisely, $A \square B$ is the set of all configurations ω for which there exist finite and disjoint sets of bonds K_A and K_B such that any configuration ω' with $\omega'(e) = 1$, for all $e \in K_A$, is in A, and any configuration ω'' with $\omega''(e) = 1$, for all $e \in K_B$, is in B.

Theorem 1.5 (BK inequality)   *Let A and B be two increasing events which depend on the state of only finitely many bonds. Then*

$$P_p(A \,\square\, B) \le P_p(A)P_p(B).$$

The requirement that A and B are allowed to depend on only finitely many bonds has a technical reason. As we shall see, this will not be important in applications. In the next chapter, we shall derive continuum analogues of Theorems 1.4 and 1.5.

Next, we discuss a method for estimating the rate of change of $P_p(A)$ as a function of p, for increasing events A. Again, we assume that A depends on the state of finitely many edges only. For this, we need yet another definition. A bond e is said to be *pivotal* for A if $1_A(\omega) \ne 1_A(\omega_e)$, where ω_e is the configuration obtained from ω by changing the value at e; i.e. $\omega_e(e) = 1 - \omega(e)$. In words, a bond is pivotal for A if the occurrence or non-occurrence of A depends crucially on the state of the bond e. It is intuitively clear that the rate of change of $P_p(A)$ as a function of p is related to the number of pivotal bonds.

Theorem 1.6 (Russo's formula)   *Let A be an increasing event which depends on only finitely many bonds. Then*

$$\frac{d}{dp} P_p(A) = \sum_{e \in \mathbf{E}} P_p(e \text{ is pivotal for } A).$$

In the discussion so far, we assumed that bonds were either open or closed with certain probabilities. There was no randomness in the vertices at all. This is the reason that we call this model *bond percolation*. But we could as well declare vertices instead of edges to be open or closed with probability p and $1 - p$ respectively, obtaining a *site* model. The discussion of this site model is similar to the discussion of the bond model above. All results in this section have a natural analogue in the site setting, although the value of the critical probability for independent site percolation on the two-dimensional integer lattice is not known. We shall use these results in the site setting freely with a possible reference to the result in the bond setting.

As mentioned before, discretisation is an important technique in the theory of continuum percolation. Sometimes we end up with a more complicated discrete lattice structure than the nearest-neighbour integer lattice. Also it might be the case that there are different types of sites which are open with different probabilities. Take the d-dimensional integer lattice and draw an edge between any two vertices v and w for which $|v - w| \le 2L$, where L is some positive constant. The graph obtained this way is denoted by $\mathcal{G}_L$. We can perform independent site percolation on this new graph, and this leads to the critical values $p_c(\mathcal{G}_L)$, $p_T(\mathcal{G}_L)$ and $p_S(\mathcal{G}_L)$.

This site-percolation model can also be extended to a multi-parametric setting. For example, consider a 'two-layered graph' $\mathcal{G}_{(L_1,L_2)}$ which is defined as follows. We place a copy of $\mathcal{G}_{L_2}$ 'above' $\mathcal{G}_{L_1}$ and we draw an edge between $v \in \mathcal{G}_{L_1}$ and $w \in \mathcal{G}_{L_2}$ if $d(v, w) \leq L_1 + L_2$ (here we abuse notation: v and w are viewed as elements of $\mathbb{R}^d$). Thus a vertex $v \in \mathcal{G}_{L_1}$ and a vertex $w \in \mathcal{G}_{L_2}$ are adjacent if and only if $d(v, w) \leq L_1 + L_2$. Now we perform multi-parameter independent site percolation on $\mathcal{G}_{(L_1,L_2)}$ by declaring a site in $\mathcal{G}_{L_1}$ to be open with probability p_1 and a site in $\mathcal{G}_{L_2}$ to be open with probability p_2. Rather than a critical point, in this model we can define a region inside the unit square where percolation occurs:

$$p_c(\mathcal{G}_{(L_1,L_2)}) = \{(p_1, p_2) \ : \ P_{(p_1,p_2)}(C(0) = \infty) > 0\},$$

where $C(0)$ is the union of the open clusters of the origins in $\mathcal{G}_{L_1}$ and $\mathcal{G}_{L_2}$. The regions $p_T(\mathcal{G}_{(L_1,L_2)})$ and $p_S(\mathcal{G}_{(L_1,L_2)})$ are defined similarly. The result which we shall need is a generalisation of Theorem 1.2:

Theorem 1.7   *In the setting just described it is the case that* $p_c(\mathcal{G}_L) = p_T(\mathcal{G}_L)$ *= $p_S(\mathcal{G}_L)$, and $p_c(\mathcal{G}_{(L_1,L_2)}) = p_T(\mathcal{G}_{(L_1,L_2)}) = p_S(\mathcal{G}_{(L_1,L_2)})$.*

We end this section with a short discussion on mixed bond/site models. In such a model, both the sites and bonds of the integer lattice are either open or closed with certain probabilities. In its most general form, we have a parameter p and for each bond or vertex, w say, there is a non-decreasing function f_w such that w is open with probability $f_w(p)$, independently of all other bonds and vertices. Many of the results quoted thus far have their analogues in the mixed setting. In particular, Theorem 1.2 is still true in this setting. Here is a version of Russo's formula for this particular setting which we shall need in Chapter 6:

Theorem 1.8 (Russo's formula)   *Consider a mixed bond/site model and let A be an increasing event which depends on the state of only finitely many vertices and bonds. Suppose in addition that there are non-decreasing differentiable functions f_b, such that the bond or vertex b is open with probability $f_b(p)$ independently of all other vertices and bonds. Then*

$$\frac{d}{dp}P_p(A) = \sum_{e \in E} P_p(e \text{ is pivotal for } A)\frac{d}{dp}f_e(p)$$

$$+ \sum_{v \in \mathbb{Z}^d} P_p(v \text{ is pivotal for } A)\frac{d}{dp}f_v(p).$$

1.3 Stationary point processes

In the discrete percolation model of the previous section, the vertices of the random graph under consideration were non-random; they were formed by the elements of the d-dimensional integer lattice. In models for continuum percolation, this is no longer the case. The positions of the vertices themselves are random, and they are formed by the occurrences of a *stationary point process*. In this section, we introduce point processes, derive some basic properties and give some examples.

One can think of a point process as a random set of points in space. But of course, this is not a very mathematical definition, and we have to make precise what we mean by 'random' here. A natural way to do this is the following. Denote the σ-algebra of Borel sets in $\mathbb{R}^d$ by $\mathcal{B}^d$, and denote by N the set of all counting measures on $\mathcal{B}^d$ which assign finite measure to bounded Borel sets and for which the measure of a point is at most 1. In this way, N can be identified with the set of all configurations of points in $\mathbb{R}^d$ without limit points. We equip N with the σ-algebra $\mathcal{N}$ generated by sets of the form

$$\{n \in N : n(A) = k\},$$

where $A \in \mathcal{B}^d$ and k is an integer. A point process can now be defined as follows:

Definition 1.4 *A point process X is a measurable mapping from a probability space $(\Omega, \mathcal{F}, P)$ into $(N, \mathcal{N})$.*

The *distribution* of X is the measure μ on $\mathcal{N}$ induced by X; i.e. μ is defined through the equation $\mu(G) = P(X^{-1}(G))$, for all $G \in \mathcal{N}$. The definition of $\mathcal{N}$ allows us to count the number of points in a set $A \in \mathcal{B}^d$: the mapping $f_A : N \to N$ defined by $f_A(n) = n(A)$ is measurable by the very construction of $\mathcal{N}$. Hence the composition $f_A \circ X : \Omega \to N$ is a random variable which we denote by $X(A)$. In words, $X(A)$ represents the random number of points inside A.

In continuum models, we do not have a nice periodic structure as in discrete percolation models. The requirement that the lattice in discrete percolation is periodic is replaced by the requirement that the point process X is *stationary*. Let T_t be the translation in $\mathbb{R}^d$ over the vector t: $T_t(s) = t + s$, for all $s \in \mathbb{R}^d$. Then T_t induces a transformation $S_t : N \to N$ through the equation

$$(S_t n)(A) = n(T_t^{-1}(A)), \tag{1.1}$$

for all $A \in \mathcal{B}^d$. On a higher level, S_t induces a transformation $\tilde{S}_t$ on measures

μ on $\mathcal{N}$ through the equation

$$(\tilde{S}_t\mu)(G) = \mu(S_t^{-1}G), \qquad (1.2)$$

for all $G \in \mathcal{N}$. Now we can define stationarity:

Definition 1.5 *The point process X is said to be* stationary *if its distribution is invariant under $\tilde{S}_t$ for all $t \in \mathbb{R}^d$.*

Definition 1.6 *The finite-dimensional (fidi) distributions of a point process X are the joint distributions, for all finite families of bounded Borel sets $A_1, \ldots, A_k$, of the random variables $X(A_1), \ldots, X(A_k)$.*

Standard methods (see e.g. Daley and Vere-Jones 1988) show that the distribution of a point process X is completely determined by its fidi distributions. The fidi distributions are thus one way of specifying a point process. In Chapter 7, we shall introduce a completely different way of specifying a stationary point process, namely via *cutting and stacking*. For now, we just note that a point process X with distribution μ is stationary if and only if the fidi distributions of μ coincide with the fidi distributions of $\tilde{S}_t(\mu)$, for all $t \in \mathbb{R}^d$.

Percolation theory is concerned with infinite objects and hence only makes sense on infinite graphs. We require therefore that our percolation models are based on point processes with the property that $X(\mathbb{R}^d) = \infty$. This, however, basically is a consequence of stationarity as we now show.

Proposition 1.1 *Let X be a stationary point process for which*

$$P(X(\mathbb{R}^d) = 0) = 0.$$

Then $P(X(\mathbb{R}^d) = \infty) = 1$.

Proof Suppose that there exists an integer k such that $P(X(\mathbb{R}^d) = k) > 0$. Then there must also exist an integer b such that

$$P\left(X(B_b) > \tfrac{1}{2}k, X(\mathbb{R}^d \setminus B_b) < \tfrac{1}{2}k\right) =: \epsilon > 0,$$

where B_b is the set $[-b, b]^d$. Let $r \in Z^d$ be a vector with integer-valued coordinates and let $br = (br_1, \ldots, br_d)$. Consider the events

$$E_r = \left\{X(T_{br}(B_b)) > \tfrac{1}{2}k, X(T_{br}(\mathbb{R}^d \setminus B_b)) < \tfrac{1}{2}k\right\}.$$

It follows from the stationarity of X that $P(E_r) = \epsilon$, for all $r \in Z^d$. But the events E_r are disjoint for distinct r, and this is the required contradiction. $\square$

Before giving some examples of point processes, we state a few definitions.

Definition 1.7 *The density of a stationary point process X is defined as $E(X([0, 1]^d))$, where E is the expectation operator corresponding to P.*

Definition 1.8 *Two point processes X_1 and X_2 defined on the same probability space are said to be* independent *if*

$$P(X_1(A_1) = k_1, \ldots, X_1(A_n) = k_n, X_2(B_1) = l_1, \ldots, X_2(B_m) = l_m)$$
$$= P(X_1(A_1) = k_1, \ldots, X_1(A_n) = k_n)$$
$$\times P(X_2(B_1) = l_1, \ldots, X_2(B_m) = l_m),$$

for every $n, m \geq 1$, l_i, k_j non-negative integers and Borel sets A_i and B_i.

Definition 1.9 *The* superposition *of two point processes X_1 and X_2 defined on the same probability space $(\Omega, \mathcal{F}, P)$ is the point process X defined by*

$$X(A)(\omega) = X_1(A)(\omega) + X_2(A)(\omega),$$

for all Borel sets A. We write $X = X_1 * X_2$ or $\mu = \mu_1 * \mu_2$, where μ, μ_1 and μ_2 are the distributions of X, X_1 and X_2 respectively.

Next we give some examples of point processes.

Example 1.1 Let U be a random d-dimensional vector defined on $(\Omega, \mathcal{F}, P)$ which is uniformly distributed in $[0, 1]^d$. Identifying a counting measure μ with the set $\{x \in \mathbb{R}^d : \mu(\{x\}) = 1\}$, we define a point process X via the equation $X(\omega) = U(\omega) + Z^d$. Hence we just shift the d-dimensional integer lattice over a random vector, and it is not hard to see, using fidi distributions, that X is stationary.

Example 1.2 (The Poisson process) The point process X is said to be a Poisson process with *density* $\lambda > 0$ if (i) and (ii) below are satisfied:

(i) For mutually disjoint Borel sets $A_1, \ldots, A_k$, the random variables $X(A_1), \ldots, X(A_k)$ are mutually independent.

(ii) For any bounded Borel set $A \in \mathcal{B}^d$ we have for every $k \geq 0$

$$P(X(A) = k) = e^{-\lambda \ell(A)} \frac{\lambda^k \ell(A)^k}{k!}, \tag{1.3}$$

where $\ell(\cdot)$ denotes Lebesgue measure in $\mathbb{R}^d$.

Note that we specified the distribution of a Poisson process by its fidi distributions. Condition (ii) guarantees that a Poisson process is stationary. Also, we have that $E(X([0, 1]^d)) = \lambda$, and hence the fact that we called λ the density of the process is consistent with Definition 1.7.

Example 1.3 (The non-homogeneous Poisson process)   The point process X is said to be a non-homogeneous Poisson process if (i) and (ii) below are satisfied:

    (i) For mutually disjoint Borel sets $A_1, \ldots, A_k$, the random variables $X(A_1), \ldots, X(A_k)$ are mutually independent.

    (ii) There exists a measurable function $\Lambda : I\!\!R^d \rightarrow [0, \infty)$, the *intensity function* of the process, such that for any bounded Borel set $A \in \mathcal{B}^d$ we have

$$P(X(A) = k) = e^{-\int_A \Lambda(x)\,dx} \frac{(\int_A \Lambda(x)\,dx)^k}{k!}. \tag{1.4}$$

We obtain a Poisson process by taking $\Lambda(x) \equiv \lambda$. Clearly, this is the only case in which we obtain a stationary process here.

We remark that condition (i) in Example 1.2 and Example 1.3 is in fact redundant. (This is a result of Renyi; see Daley and Vere-Jones 1988, Theorem 2.3.1.) In this book, however, we use the independence property frequently and that is the reason to highlight condition (i) in both examples.

Suppose that we have a sequence of bounded Borel sets $A_n \subset I\!\!R^d$ which increases to A, where we do not require A to be bounded. Obviously, the events $\{X(A_n) \geq k\}$ increase to $\{X(A) \geq k\}$ when $n \rightarrow \infty$, and hence $P(X(A_n) = k) \rightarrow P(X(A) = k)$. It follows from monotone convergence that

$$
\begin{aligned}
P(X(A) = k) &= \lim_{n \to \infty} P(X(A_n) = k) \\
&= \lim_{n \to \infty} e^{-\int_{A_n} \Lambda(x)\,dx} \frac{(\int_{A_n} \Lambda(x)\,dx)^k}{k!} \\
&= e^{-\int_A \Lambda(x)\,dx} \frac{(\int_A \Lambda(x)\,dx)^k}{k!},
\end{aligned}
$$

where the last expression is to be interpreted as zero when $\int_A \Lambda(x)\,dx = \infty$.

We end this section with some properties of the Poisson process. This process is of particular interest, and four chapters of this book are concerned with percolation models based on a Poisson process.

In the physics literature, people often use phrases like 'consider infinitely many points uniformly distributed in space'. In fact, they refer to a Poisson process in such a case, and the property they isolate in that phrase should be interpreted as in the following proposition.

Proposition 1.2 *Let X be a Poisson process and A be a Borel set with bounded positive Lebesgue measure. Then, for all measurable B ⊂ A, we have*

$$P(X(B) = m \mid X(A) = m + k) = \binom{m + k}{m} \left(\frac{\ell(B)}{\ell(A)}\right)^m \left(1 - \frac{\ell(B)}{\ell(A)}\right)^k.$$

Proof This follows from straightforward calculations.   □

Another useful property of the Poisson process is the following: suppose we condition on the event that there is a point at x, for some $x \in \mathbb{R}^d$. The independence property of the Poisson process now implies that, apart from the given point at x, the probabilistic structure of the conditioned process is identical with that of the original process. Writing μ for the distribution of our Poisson process, μ_x for the process conditioned to have a point at x, and δ_x for the distribution of an independent process having only one point at x a.s., we can write this property as

$$\mu_x = \mu * \delta_x. \tag{1.5}$$

The distribution μ_x is called the *Palm distribution* of μ. For the tedious technical details, we refer to Daley and Vere-Jones (1988). If we condition on the event that there is a point at the origin, we still obtain, apart from that point at the origin, a Poisson process.

The third property we discuss is that the superposition of two independent Poisson processes X_{λ_1} and X_{λ_2} with density λ_1 and λ_2, respectively, is again a Poisson process with density $\lambda_1 + \lambda_2$. The reason for this is that if we take the sum of two independent random variables with Poisson distribution with parameters λ_1 and λ_1, respectively, we obtain a random variable with Poisson distribution with parameter $\lambda_1 + \lambda_2$. Hence, in the obvious notation:

$$X_{\lambda_1} * X_{\lambda_2} = X_{\lambda_1 + \lambda_2}. \tag{1.6}$$

Finally, we show how one can obtain a non-homogeneous Poisson process from an ordinary Poisson process in a probabilistic way. We assume that the probability space is rich enough for our purposes here. Let X be a Poisson process with density λ, and let $g : \mathbb{R}^d \to [0, 1]$ be a measurable mapping. We consider a realisation $X(\omega)$ of X. If there is a point at x, we take the point away with probability $1 - g(x)$ and leave it where it is with probability $g(x)$, independently of all other points of the Poisson process. The ensuing point process is denoted by $\tilde{X}$. Thus, $\tilde{X}$ is a *thinning* of the original process X.

Proposition 1.3 *The point process $\tilde{X}$ is a non-homogeneous Poisson process with intensity function λg.*

Proof The independence property is immediate from the construction. The fidi distribution of $\tilde{X}$ can be computed as follows:

$$P(\tilde{X}(A) = k) = \sum_{i=k}^{\infty} P(X(A) = i)P(\tilde{X}(A) = k \mid X(A) = i).$$

We have from Proposition 1.2 that given the event $\{X(A) = i\}$, the i points of X in A are uniformly distributed over A. Thus

$$P(\tilde{X}(A) = 1 \mid X(A) = 1) = \ell(A)^{-1} \int_A g(x)\, dx,$$

and more generally,

$$P(\tilde{X}(A) = k \mid X(A) = i) = \binom{i}{k}(\ell(A)^{-1} \int_A g(x)\, dx)^k [1 - \ell(A)^{-1}$$
$$\times \int_A g(x)\, dx]^{i-k}.$$

Hence

$$P(\tilde{X}(A) = k) = e^{-\lambda \ell(A)} \frac{(\lambda \int_A g(x)\, dx)^k}{k!} \times$$
$$\times \sum_{i=k}^{\infty} \frac{(\lambda \ell(A)[1 - \ell(A)^{-1} \int_A g(x)\, dx])^{i-k}}{(i-k)!}$$
$$= e^{-\lambda \ell(A)} \frac{(\lambda \int_A g(x)\, dx)^k}{k!} e^{\lambda \ell(A)(1-\ell(A)^{-1} \int_A g(x)\, dx)}$$
$$= \frac{(\lambda \int_A g(x)\, dx)^k}{k!} e^{-\lambda \int_A g(x)\, dx}. \qquad \square$$

Note that if g is a constant function then we obtain a homogeneous Poisson process with lower density than the original process.

1.4 The Boolean model

The first model of continuum percolation which we introduce in this chapter is the Poisson blob model, or the Boolean model as we shall call it throughout the book. In this section we give the formal mathematical construction and we fix notation for this model.

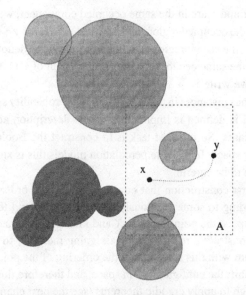

Figure 1.1. A realisation of a Boolean model; the shaded region is C, the darker shaded region is W, while V is empty here. Finally, $x \overset{\vee}{\leadsto} y$ in A.

Before giving the mathematical details, let us try to say in words what this model is all about. We start with some stationary point process X, as introduced in the previous section. We say that the model is *driven* by X. Each point of X is the centre of a closed ball (in the usual Euclidean metric) with a random radius in such a way that radii corresponding to different points are independent of each other and identically distributed. The radii are also independent of X. In this way, the space is partitioned into two regions, the *occupied* region, which is the region covered by at least one ball, and the *vacant* region, which is just the complement of the occupied region. (See Figure 1.1.) The occupied region is denoted by C. Both the occupied and vacant regions consist of connected components, and almost all results in this book have to do with these components. The connected components in the occupied region will be called *occupied components*. Similarly, the connected components in the vacant region are called *vacant components*. For $A \subset \mathbb{R}^d$, we denote by $W(A)$ the union of all occupied components which have non-empty intersection with A. When $A = \{0\}$, we write $W := W(\{0\})$. We call W the occupied component of the origin. In the case of vacancy, all definitions are similar, using the symbol V instead of W. We say that V is the vacant component of the origin. Note that either V or W is empty but not both. The ball centred at x is denoted by $S(x)$ or by $S(x, r)$, where r denotes the (random) radius of the ball.

If two points x and y are in the same occupied component, we say that they are *connected* in the occupied region, and we write $x \overset{o}{\leadsto} y$ (or $y \overset{o}{\leadsto} x$ of course). Connectedness in the vacant region is defined similarly, and denoted by $x \overset{v}{\leadsto} y$. If x and y are in the same occupied (vacant) component of $C \cap A$ ($C^c \cap A$) for some $A \subset \mathbb{R}^d$, we write $x \overset{o}{\leadsto} y$ in A ($x \overset{v}{\leadsto} y$ in A).

There are various instances in this book where the probability space on which the whole model is defined is important, and the description above does not suffice in such cases. So our first task is to construct the Boolean model on some probability space. In discrete percolation models this is straightforward, but here we have to be careful.

Perhaps the first construction that comes to mind is to order the points of X linearly according to some previously determined rule and to construct on one probability space the point process X and i.i.d. positive random variables $Y_1, Y_2, \ldots$ and construct a realisation by assigning radius Y_i to the i-th point of X. The problem with this setup is that the ordering of the points of X is not preserved if we shift the configuration in space, and therefore this construction is not good enough to apply ergodic theorems (see the next chapter).

Another approach is to consider uncountably many random variables $\{Y_x\}$ indexed by $\mathbb{R}^d$ and assign radius Y_x to the point of X at x if it exists. The problem with this setup is that even the simplest events will not be measurable.

We shall now describe a construction based on only countably many random variables in such a way that we can define shifts. Let X be defined on some probability space $(\Omega_1, \mathcal{F}_1, P_1)$. Let Ω_2 be the product space $\prod_{n \in N} \prod_{z \in \mathbb{Z}^d} [0, \infty)$ and equip Ω_2 with the usual product σ-field and product measure P_2 with all marginals being μ, where μ is a probability measure on $[0, \infty)$. An element $\omega_2 \in \Omega_2$ is sometimes denoted by $\omega_2(n, z)$. Finally, we set $\Omega = \Omega_1 \times \Omega_2$ and equip Ω with product measure $P = P_1 \times P_2$ and the usual product σ-algebra. A Boolean model is a measurable mapping from Ω into $N \times \Omega_2$ defined by $(\omega_1, \omega_2) \to (X(\omega_1), \omega_2)$, where N is as defined in Section 1.3. The configuration of balls in space corresponding to (ω_1, ω_2) is obtained as follows. Consider *binary cubes*

$$K(n, z) := \prod_{i=1}^{d} (z_i 2^{-n}, (z_i + 1)2^{-n}] \quad \text{for all } n \in N \text{ and } z \in \mathbb{Z}^d.$$

We call this a binary cube of order n. Each point $x \in X$ is contained in a unique binary cube of order n, $K(n, z(n, x))$ say, and with P_1-probability 1, for each point $x \in X$ there is a unique smallest number $n_0 = n_0(x)$ such that $K(n_0, z(n_0, x))$ contains no other points of X. The radius of the ball centred at x is now defined to be $\omega_2(n_0, z(n_0, x))$.

The product structure of Ω implies that the radii are independent of the point process, and the product structure of Ω_2 implies that different points have balls with independent, identically distributed radii. It is natural to denote this model by (X, μ). In most cases, however, we have a certain random variable ρ with distribution μ and we think of this random variable as governing the radii of the balls: the radii of the balls are random and are distributed as ρ. The model is then denoted by (X, ρ). In the case where X is a Poisson process with density λ we shall write $P = P_\lambda = P_{(\lambda, \rho)}$ to emphasise the dependence on the parameter. Also, the probability of an event A is denoted by either $P(A)$ or $P\{A\}$ depending on the circumstances.

Let the unit vectors in $\mathbb{R}^d$ be denoted by $e_1, \ldots, e_d$. The translation $T_{e_i} : \mathbb{R}^d \to \mathbb{R}^d$ defined by $x \to x + e_i$ from the previous section induces a transformation U_{e_i} on Ω_2 through the equation

$$(U_{e_i}\omega_2)(n, z) = \omega_2(n, z - e_i). \tag{1.7}$$

As before, S_{e_i} is defined on Ω_1 via the equation

$$(S_{e_i}\omega_1)(A) = \omega_1(T_{e_i}^{-1}A). \tag{1.8}$$

Hence T_{e_i} induces a transformation $\tilde{T}_{e_i}$ on $\Omega = \Omega_1 \times \Omega_2$ defined by

$$\tilde{T}_{e_i}(\omega) = (S_{e_i}\omega_1, U_{e_i}\omega_2). \tag{1.9}$$

The transformation $\tilde{T}_{e_i}$ corresponds to a translation by the vector e_i of a configuration of balls in space. As such, it will play a crucial role in the discussion of ergodicity in the next chapter.

In percolation theory, one is mainly interested in unbounded objects. In the present setting, this means that we are interested in unbounded occupied and vacant components. The most basic question one can ask about unbounded components concerns their existence. Given a certain Boolean model (X, ρ), is there a positive probability that the occupied or vacant component of the origin is unbounded? (The fact that we take the origin here is of no importance. From the stationarity of X and the independence of the radii, we cannot distinguish between different points of the space probabilistically.) For a given Boolean model (X, ρ), this question is usually very hard to answer. Instead, one considers a whole family of Boolean models and then proves that certain members of the family do not allow unbounded components, but others do. To whet the reader's appetite, let us look as an example at occupied components in Boolean models driven by Poisson processes. A Boolean model driven by a Poisson process with density λ and radius random variable ρ is denoted by (X, ρ, λ), and we call it a *Poisson Boolean model*. We denote by $\theta_\rho(\lambda) = \theta(\lambda)$ the probability that the origin is an element of an unbounded occupied component. In other words, if

$d(A)$ denotes the diameter of a set $A \subset I\!R^d$ (i.e. $d(A) = \sup_{x,y \in A} |x - y|$), $\theta(\lambda)$ is the probability that $d(W) = \infty$. The function θ is called the *percolation function*. It *seems* obvious (but try to prove this!) that θ is non-decreasing in λ, and for the moment we assume this. A rigorous proof of this fact using coupling is given in Section 2.2. We can define the *critical density* $\lambda_c = \lambda_c(\rho)$ as follows:

$$\lambda_c(\rho) = \inf\{\lambda \geq 0 : \theta_\rho(\lambda) > 0\}. \tag{1.10}$$

In Chapter 3, we shall prove that λ_c is *non-trivial* in all reasonable cases; i.e. λ_c is strictly positive and finite. This fact is at the heart of percolation theory, and it immediately implies that for $\lambda > \lambda_c$, unbounded occupied components do indeed exist with positive probability. For $\lambda < \lambda_c$, the origin has probability zero to be contained in an unbounded occupied component and it follows immediately from the stationarity of the process that so has any other point. But any unbounded occupied component should contain at least one point with rational coordinates of which there are only countably many. Hence, no unbounded occupied components can exist with positive probability for $\lambda < \lambda_c$. When $\lambda < \lambda_c$, we say that the system is in the *subcritical phase*; when $\lambda > \lambda_c$, the system is said to be *supercritical*. At the critical density λ_c the system is said to be *critical*.

We can define critical densities for unbounded vacant components in a similar manner. We write $\theta_\rho^*(\lambda)$ for the probability that $d(V) = \infty$ and the critical density $\lambda_c^* = \lambda_c^*(\rho)$ is defined as

$$\lambda_c^*(\rho) = \sup\{\lambda \geq 0 : \theta_\rho^*(\lambda) > 0\}. \tag{1.11}$$

In Chapters 3, 4 and 5 the Poisson Boolean model is studied extensively. Boolean models driven by general point processes are treated in Chapter 7.

1.5 The random-connection model

Given a stationary point process X, there is another natural way of constructing unbounded random objects and this section is concerned with this second model. Again, we first introduce the model in an informal way.

In a Boolean model, the second characteristic of the model (the first is of course the point process X) is a random variable ρ which governs the behaviour of the radii of the balls. In a random-connection model (RCM), the second characteristic is a so-called *connection function*, which is a non-increasing function from the positive reals into $[0, 1]$. Given a connection function g, the rule is as follows: for any two points x_1 and x_2 of the point process X, we insert an edge between x_1 and x_2 with probability $g(|x_1 - x_2|)$, independently of all other pairs of points of X, where $|\cdot|$ denotes the usual Euclidean distance. The edge between

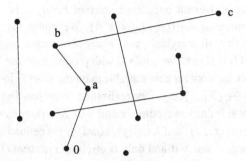

Figure 1.2. A realisation of a random-connection model. Here, W contains the points $\{0, a, b, c\}$.

two points x_1 and x_2 is denoted by the unordered pair $\{x_1, x_2\}$, and we say that x_1 and x_2 are the *end vertices* of $\{x_1, x_2\}$. Two points x and y of the process are said to be *connected* if there exists a finite sequence $(x =: x_1, x_2, \ldots, x_n := y)$ such that the edge $\{x_i, x_{i+1}\}$ is inserted for all $i = 1, \ldots, n - 1$. A *component* can now be defined in the usual graph-theoretical way: a component is a set of points such that any two points of this set are connected to each other, and which is maximal with respect to this property. The occupied component of the origin is denoted by W. Of course for W to be non-empty, we need to condition the process to have a point at the origin (see Figure 1.2). This is the natural analogue of the occupied components in Boolean models. There is no analogue of vacant components in random-connection models. We again say that the RCM is *driven* by X, and the model is denoted by (X, g).

We remark that ordinary nearest-neighbour bond percolation on $\mathbf{Z}^d$ is a special case of a random-connection model. To see this, just take the point process of Example 1.1 in Section 1.3; i.e. we shift the d-dimensional integer lattice over a random vector. For the connection function g we can take $g(x) = p1_{\{|x| \leq 1\}}$. In this sense, a random-connection model is more general than ordinary discrete percolation.

Here is a formal mathematical construction of a random-connection model (X, g); it is quite similar to the one of a Boolean model. The notation will be basically the same as in the Boolean model, but this will not cause any confusion, as it is always clear which model is under consideration. First we assume that the point process X is defined on a probability space $(\Omega_1, \mathcal{F}_1, P_1)$. Next we consider a second probability space Ω_2 defined as

$$\Omega_2 = \prod_{\{K(n,z), K(m,z')\}} [0, 1],$$

where the product is over all unordered pairs of binary cubes. An element $\omega_2 \in \Omega_2$ is written as $\omega_2(\{(n, z), (m, z')\})$. We equip Ω_2 with product measure P_2 such that all marginals are Lebesgue measure on $[0, 1]$. As before, we set $\Omega = \Omega_1 \times \Omega_2$ and we equip Ω with product measure $P = P_1 \times P_2$. A random-connection model is a measurable mapping from Ω into $N \times \Omega_2$ defined by $(\omega_1, \omega_2) \to (X(\omega_1), \omega_2)$. The realisation corresponding to (ω_1, ω_2) is obtained as follows: for any two points x and y of $X(\omega_1)$, consider the binary cubes $K(n_0(x), z(n_0(x), x))$ and $K(n_0(y), z(n_0(y), y))$ defined in the previous section. We connect x and y if and only if $\omega_2(\{(n_0(x), z(n_0(x), x)), (n_0(y), z(n_0(y), y))\}) < g(|x - y|)$. The transformations U_{e_i} on Ω_2 and $\tilde{T}_{e_i}$ on Ω can now be defined similar to (1.7) and (1.9) respectively. The transformation $\tilde{T}_{e_i}$ again corresponds to shifting a realisation of the RCM by the vector e_i in space.

1.6 Notes

Continuum percolation models have been extensively studied by physicists. Most of their results are based on Monte Carlo simulations supported by heuristic arguments. The non-triviality of the critical probabilities in Theorem 1.1 goes back to Broadbent and Hammersley (1957). Theorem 1.2 was proved independently by Menshikov (1986) and Aizenman and Barsky (1987). The fact that $p_c(2) = \frac{1}{2}$ is due to Kesten (1980). The uniqueness of the infinite cluster was proved by Harris (1960) in two dimensions and by Aizenman, Kesten and Newman (1987) in all dimensions. See Meester (1994) for a review on uniqueness in percolation theory. The FKG inequality goes back to Fortuin, Kasteleyn and Ginibre (1971), and the BK inequality was obtained for increasing events by v.d. Berg and Kesten (1985) and in general by Reimer (1994). Russo's formula is due to Russo (1978). It seems that continuum percolation models appeared for the first time in Gilbert (1961), in a very applied fashion.

2

Basic methods

In this chapter we present a few basic results which will be used repeatedly in the subsequent development of the subject. The last few sections of this chapter will be devoted to obtain some inequalities, which are the continuum version of similar theorems in the discrete percolation models. Some proofs here will be obtained by a suitable discretisation and approximation, thus we will be making use of the corresponding results in discrete percolation. The first two sections are devoted to the concepts of ergodicity, coupling and scaling. The techniques of the proofs are hardly needed in the rest of the book, so it is quite possible to read the statements of the main results and move on to the next chapter.

2.1 Ergodicity

In this section, we review some results from classical ergodic theory and apply this theory to stationary point processes. The account on ergodic theory will be fairly short; we restrict ourselves to those results which we need in this book. More information about ergodicity and stationary point processes can be found in the book of Daley and Vere-Jones (1988). For a general account on ergodic theory, we refer to the books by Krengel (1985) and Petersen (1983). It will be very convenient here to use a slightly different notation than in the rest of the book in order to clearly see stationary point processes from the viewpoint of measure-preserving transformations (which we introduce in the next paragraph).

Consider a probability space $(\Omega, \mathcal{F}, \mu)$ and let $T : \Omega \to \Omega$ be an invertible *measure-preserving (m.p.) transformation*, that is, $\mu(T^{-1}F) = \mu(F)$, for all $F \in \mathcal{F}$. We call the quadruple $(\Omega, \mathcal{F}, \mu, T)$ an *m.p. dynamical system*. An element $F \in \mathcal{F}$ is said to be *T-invariant* if $T^{-1}F = F$. Clearly, the set of all

T-invariant sets in $\mathcal{F}$ forms a σ-algebra, which we denote by $\mathcal{I}$. The classical one-dimensional ergodic theorem is as follows:

Proposition 2.1   *Let $(\Omega, \mathcal{F}, \mu, T)$ be an m.p. dynamical system and let f be a μ-integrable real function on Ω. Then*

$$\frac{1}{n} \sum_{i=0}^{n-1} f(T^i(\omega)) \to E(f \,|\, \mathcal{I})(\omega), \tag{2.1}$$

μ-a.s. when $n \to \infty$.

To apply this proposition to stationary one-dimensional point processes, it is convenient to identify the space $(N, \mathcal{N})$ with $(\Omega, \mathcal{F})$, so that any element $\omega \in \Omega$ represents a counting measure in $\mathbb{R}^d$ and any measure μ on $\mathcal{F}$ is identified with a stationary point process. Let T_t be the shift by a distance t to the right in $\mathbb{R}$. Then T_t induces a transformation $S_t : \Omega \to \Omega$ through the equation

$$(S_t \omega)(A) = \omega(T_t^{-1} A), \tag{2.2}$$

where A is a measurable subset of $\mathbb{R}$.

Lemma 2.1   *If μ is stationary, then S_t is measure preserving, for all $t \in \mathbb{R}$.*

Proof   For a bounded measurable set $A \subset \mathbb{R}$ and $k \in N$, let F be the set $\{\omega : \omega(A) = k\} \in \mathcal{F}$. Then we have $\mu(F) = \mu\{\omega : \omega(A) = k\} = \mu\{\omega : \omega(T_t^{-1}(A)) = k\}$ by stationarity of μ. Using (2.2), this is equal to $\mu\{\omega : (S_t \omega)(A) = k\} = \mu(S_t^{-1} F)$. The desired equality is then also true for sets of the form $\cap_{i=1}^{n} \{\omega : \omega(A_i) = k_i\}$ where A_i is bounded and measurable for all i. Hence we have shown the necessary equality for a generating π-system of $\mathcal{F}$ and the proof is complete. $\square$

Taking $t = -1$ for convenience, we now have an m.p. dynamical system $(\Omega, \mathcal{F}, \mu, S_{-1})$ and we can apply (2.1) as follows. Consider a point process with finite density and let $f(\omega) := \omega(0, 1]$. Obviously, we have $f(S_{-1}^i(\omega)) = \omega(i, i+1]$, and hence it follows that $\sum_{i=0}^{n-1} f(S_{-1}^i(\omega)) = \omega(0, n]$. It follows from (2.1) that

$$n^{-1} \omega(0, n] \to E(f \,|\, \mathcal{I}_1)(\omega), \text{ a.s.} \tag{2.3}$$

for $n \to \infty$, where $\mathcal{I}_1$ is the σ-algebra of S_{-1}-invariant events. Note that n is an integer here. We can strengthen the conclusion to arbitrary intervals using the following inequalities, where $\lfloor t \rfloor$ denotes the largest integer smaller than or equal to t:

$$\frac{\lfloor t \rfloor}{t} \cdot \frac{\omega(0, \lfloor t \rfloor]}{\lfloor t \rfloor} \leq \frac{\omega(0, t]}{t} \leq \frac{\lfloor t+1 \rfloor}{t} \cdot \frac{\omega(0, \lfloor t+1 \rfloor]}{\lfloor t+1 \rfloor}.$$

We conclude that

$$t^{-1}\omega(0, t] \to E(f \,|\, \mathcal{I}_1)(\omega) \text{ a.s.} \tag{2.4}$$

when $t \to \infty$.

Before we continue with the higher-dimensional case, we make the following remark. Let $\mathcal{I}$ be the σ-algebra of events which are invariant under *all* translations S_t, $t \in \mathbb{R}$. Of course we have that $\mathcal{I} \subset \mathcal{I}_1$. Now note that for a fixed positive number t_0 we have a.s.

$$\lim_{t \to \infty} t^{-1}(S_{t_0}\omega)(0, t] = \lim_{t \to \infty} t^{-1}(\omega(-t_0, 0] + \omega(0, t - t_0])$$
$$= \lim_{t \to \infty} t^{-1}\omega(0, t],$$

so that the right-hand side in (2.4) is invariant under translations and hence measurable with respect to $\mathcal{I}$. But then we have a.s.

$$E(f \,|\, \mathcal{I}_1) = E(E(f \,|\, \mathcal{I}_1) \,|\, \mathcal{I})$$
$$= E(f \,|\, \mathcal{I}),$$

and we finally obtain the important formula

$$t^{-1}\omega(0, t] \to E(f \,|\, \mathcal{I})(\omega) \text{ a.s.} \tag{2.5}$$

for $t \to \infty$.

Next we consider the higher-dimensional case. As before, $(\Omega, \mathcal{F}, \mu)$ is a probability space, but now we consider d invertible, commuting, measurable and m.p. transformations $T_1, \ldots, T_d$ from Ω into itself, where $d \geq 1$. Let $\mathcal{I}_1$ be the σ-field of events in $\mathcal{F}$ which are invariant under all transformations $T_1, \ldots, T_d$. The composition $T_1^{i_1} \cdots T_d^{i_d}$ is a transformation which we denote by $T_{(i_1, \ldots, i_d)}$. In this way the set $\{T_z \,:\, z \in Z^d\}$ forms a group and is called a Z^d-action on $(\Omega, \mathcal{F}, \mu)$. We also say that Z^d acts on Ω via the transformations $T_1, \ldots, T_d$. The classical higher-dimensional ergodic theorem is as follows:

Proposition 2.2 *Let $(\Omega, \mathcal{F}, \mu)$ be a probability space and suppose Z^d acts on Ω via $T_1, \ldots, T_d$ (which are supposed to be measure preserving). Let f be a real μ-integrable function on Ω. Then*

$$\frac{1}{n^d} \sum_{i_1=0}^{n-1} \cdots \sum_{i_d=0}^{n-1} f(T_1^{i_1} \cdots T_d^{i_d}(\omega)) \to E(f \,|\, \mathcal{I}_1)(\omega), \text{ a.s.}$$

when $n \to \infty$.

This proposition may now be applied to d-dimensional stationary point processes with finite density. For this, let $x \in {I\!\!R}^d$ and let T_x be the translation over the vector x. For ease of notation, we define T_i to be the shift over the i-th unit vector in ${I\!\!R}^d$ and S_x is defined by the higher-dimensional analogue of (2.2). The same reasoning as above yields the higher-dimensional analogue of (2.5):

$$t^{-d}\, \omega(0, t]^d \rightarrow E(f | \mathcal{I})(\omega) \text{ a.s.} \tag{2.6}$$

where $\mathcal{I}$ is the σ-algebra of events which are invariant under all transformations $\{S_x : x \in {I\!\!R}^d\}$, and $f(\omega) = \omega(0, 1]^d$.

Up to this point, we have been able to apply 'discrete time' ergodic theorems to obtain conclusions in continuous time like (2.5) and (2.6). Point processes are random measures in a continuum, and therefore the most natural ergodic theorems associated with them are ergodic theorems which are concerned with a group of transformations indexed by ${I\!\!R}^d$ rather than Z^d. An ${I\!\!R}^d$-action $\{S_x : x \in {I\!\!R}^d\}$ is a group of invertible, commuting, m.p. transformations acting measurably on a probability space $(\Omega, \mathcal{F}, \mu)$ and indexed by ${I\!\!R}^d$. Our final ergodic theorem is the following:

Proposition 2.3 *Let $(\Omega, \mathcal{F}, \mu)$ be a probability space and let $\{S_x : x \in {I\!\!R}^d\}$ be an ${I\!\!R}^d$-action on Ω. Let f be a real measurable and μ-integrable function on Ω. Then,*

$$\frac{1}{t^d} \int_{[0,t]^d} f(S_x(\omega))dx \rightarrow E(f | \mathcal{I})(\omega) \text{ a.s.} \tag{2.7}$$

for $t \rightarrow \infty$, where $\mathcal{I}$ is the σ-algebra of events which are invariant under the whole group $\{S_x : x \in {I\!\!R}^d\}$.

The conditional expectations which appear in all ergodic theorems above may not be so easy to deal with in general. A very important special case occurs when the σ-algebra of invariant events is trivial, i.e. any invariant event has measure either zero or one.

Definition 2.1 *An m.p. dynamical system $(\Omega, \mathcal{F}, \mu, T)$ is said to be* ergodic *if the σ-algebra of T-invariant events is trivial. An ${I\!\!R}^d$-action or Z^d-action is said to be* ergodic *(or to* act ergodically*) if the σ-algebra of events invariant under the whole group is trivial.*

In the rest of this section, T_x is the translation in ${I\!\!R}^d$ by the vector x, and S_x is the corresponding operator on the probability space $(\Omega, \mathcal{F}, \mu)$ defined via (2.2).

Definition 2.2 *A stationary point process μ is said to be ergodic if $\{S_x \ : \ x \in \mathbb{R}^d\}$ acts ergodically on $(\Omega, \mathcal{F}, \mu)$.*

If μ is an ergodic point process, then the σ-algebra of invariant events $\mathcal{I}$ in (2.6) is trivial and hence $E(f \,|\, \mathcal{I}) = Ef$ a.s. and the limit in (2.6) is an a.s. constant. When $f(\omega) = \omega(0, 1]^d$, this constant is equal to the density of the point process. Hence we can immediately write down the following proposition.

Proposition 2.4 *For a stationary ergodic point process with finite density, the average number of points per unit volume in $[0, t]^d$ converges for $t \to \infty$ a.s. to the density of the point process.*

It is not hard to show, using this proposition and a 'thinning argument', that a corresponding statement holds for infinite-density point processes. In such cases, the limit is infinity a.s.

In general it is very difficult to determine whether or not a given point process is ergodic. The following result can be of some help as it characterises ergodicity.

Proposition 2.5 *The group $\{S_x \ : \ x \in \mathbb{R}^d\}$ acts ergodically on $(\Omega, \mathcal{F}, \mu)$ if and only if, for all $E, F \in \mathcal{F}$ we have*

$$\lim_{t \to \infty} \frac{1}{t^d} \int_{[0,t]^d} \mu(S_x E \cap F)dx = \mu(E)\mu(F). \qquad (2.8)$$

Proof Suppose that (2.8) holds and let E be an invariant event. This implies that $\mu(S_x E \cap E) = \mu(E)$, and it follows from (2.8) that

$$\frac{1}{t^d} \int_{[0,t]^d} \left\{ \mu(E) - (\mu(E))^2 \right\} dx \to 0,$$

for $t \to \infty$. Thus we have $\mu(E) = (\mu(E))^2$ and we are done.

Conversely, suppose that any invariant event has measure 0 or 1. Then (2.7) takes the form

$$\frac{1}{t^d} \int_{[0,t]^d} f(S_x(\omega))dx \to Ef, \text{ a.s.} \qquad (2.9)$$

Now take $E, F \in \mathcal{F}$ and take $f(\omega) = 1_E(\omega)$ so that $Ef = \mu(E)$. Then (2.9) yields

$$\frac{1}{t^d} \int_{[0,t]^d} 1_E(S_x(\omega))dx \to \mu(E), \text{ a.s.} \qquad (2.10)$$

The assertion now follows from the following equalities, using Fubini's theorem, (2.10) and bounded convergence:

$$\lim_{t \to \infty} t^{-d} \int_{[0,t]^d} \int_F 1_E(S_x(\omega)) d\mu(\omega) dx$$

$$= \lim_{t \to \infty} \int_F t^{-d} \int_{[0,t]^d} 1_E(S_x(\omega)) dx d\mu(\omega)$$

$$= \int_F \mu(E) d\mu(\omega)$$

$$= \mu(E)\mu(F). \qquad \square$$

This proposition is often used by checking that the group $\{S_x : x \in \mathbb{R}^d\}$ satisfies an even stronger property than (2.8), namely the *mixing* property:

Definition 2.3 An m.p. dynamical system $(\Omega, \mathcal{F}, \mu, T)$ is said to be mixing if for all $E, F \in \mathcal{F}$, $\mu(T^n E \cap F) - \mu(E)\mu(F) \to 0$, for $n \to \infty$. An $\mathbb{R}^d$-action $\{S_x : x \in \mathbb{R}^d\}$ is said to be mixing if for all $E, F \in \mathcal{F}$ we have

$$\mu(S_x E \cap F) - \mu(E)\mu(F) \to 0, \qquad (2.11)$$

when $|x| \to \infty$. For a $\mathbb{Z}^d$-action the definition is similar.

It is clear from this definition and (2.8) that a mixing point process is also ergodic. It is now easy to prove the following result:

Proposition 2.6 *A Poisson point process is ergodic.*

Proof We prove the stronger statement that a Poisson process is mixing. When E and F are events which depend only on the realisation of the point process inside a bounded set, then (2.11) follows immediately because of the independence property of the Poisson process. For arbitrary events E and F, one approximates E and F by events which depend only on the realisation of the point process inside a bounded set, and the result follows easily. $\square$

Let μ be a Poisson process with parameter λ. We can now apply Proposition 2.4 to μ and conclude that

$$t^{-d} \omega(0, t]^d \to \lambda, \text{ a.s.} \qquad (2.12)$$

when $t \to \infty$. For the Poisson process, however, we do not really need the ergodic theorem to derive (2.12). The result also follows from the classical strong law of large numbers, because the number of points in disjoint sets is independent.

Consider an ergodic point process μ. By definition, this means that any event which is invariant under the whole group of transformations $\{S_x : x \in \mathbb{R}^d\}$ has measure zero or one. This fact, a priori, does not give any information about the question of whether or not a particular transformation S_{x_0} gives rise to an ergodic m.p. dynamical system $(\Omega, \mathcal{F}, \mu, S_{x_0})$. (Of course, the reverse is much simpler: if S_{x_0} is ergodic, then the σ-algebra $\mathcal{I}_{x_0}$ of S_{x_0}-invariant events is trivial. The σ-algebra of events invariant under the whole group is contained in $\mathcal{I}_{x_0}$ and hence also trivial.) Sometimes we need to find a particular ergodic transformation when the group acts ergodically. For this, we need the following classical result:

Proposition 2.7 *Suppose that $\mathbb{R}^d$ acts ergodically on a probability space $(\Omega, \mathcal{F}, \mu)$ via the group $\{S_x : x \in \mathbb{R}^d\}$, and suppose that the σ-algebra $\mathcal{F}$ is countably generated. Then there exists a countable set of hyperplanes (where a hyperplane is not assumed to contain the origin) such that for all elements $x \in \mathbb{R}^d$ which are not contained in any of these hyperplanes, the m.p. dynamical system $(\Omega, \mathcal{F}, \mu, S_x)$ is ergodic.*

Note that we can safely apply this result to point processes: the σ-algebra $\mathcal{F}$ is generated by sets of the form $\{\omega : \omega(A) = k\}$, where $A \subset \mathbb{R}^d$ is a rectangle with rational coordinates and k an integer. The corresponding problem for mixing point processes is trivial. It follows immediately from Definition 2.3 that for a mixing point process μ, any element $x \in \mathbb{R}^d$ gives rise to a mixing m.p. dynamical system.

Finally, we discuss the ergodicity of Boolean models and random-connection models. The construction of the Boolean model and the random-connection model is such that it only makes sense to consider translations by integer-valued vectors. Therefore a Boolean model is said to be ergodic if the group $\{\tilde{T}_z : z \in Z^d\}$ acts ergodically. For a random-connection model, the definition is similar. The following result shows that ergodicity of a point process carries over to a Boolean model or RCM driven by that process:

Proposition 2.8 *Suppose X is ergodic. Then any Boolean model (X, ρ) or RCM (X, g) is also ergodic.*

Proof We give the proof for the Boolean model. From Proposition 2.7 we have that there exists a $t_0 \in \mathbb{R}^d$ with $|t_0| = 1$ such that $(\Omega_1, \mathcal{F}_1, P_1, S_{t_0})$ is an ergodic m.p. dynamical system. In the construction of the Boolean model, we can rotate the coordinate axes in such a way that t_0 becomes one of the unit vectors. It is therefore no loss of generality to assume that $t_0 = e_1$. It is obvious that $(\Omega_2, \mathcal{F}_2, P_2, U_{e_1})$ is a mixing system, and it is a classical result in ergodic theory (see e.g. Petersen 1983, Theorem 6.1) that this implies that the product transformation $\tilde{T}_{e_1}$ is ergodic. Hence $(\Omega, \mathcal{F}, P, \tilde{T}_{e_1})$ is an ergodic m.p. dynamical system and it follows that $\{\tilde{T}_z : z \in \mathbf{Z}^d\}$ acts ergodically. $\square$

We end this section with an application of Proposition 2.8.

Theorem 2.1 *Suppose that the Boolean model (X, ρ) or the RCM (X, g) is driven by an ergodic point process X. In the Boolean model, the number of unbounded occupied components is a constant a.s. and the same is true for the number of unbounded vacant components. In the RCM, the number of unbounded components is a constant a.s.*

Proof   Again, we give the proof in the case of a Boolean model. Denote the (random) number of unbounded occupied (or vacant) components by N. It is clear that the event $\{N = k\}$ is invariant under the group $\{\tilde{T}_z : z \in \mathbf{Z}^d\}$, for all $k \geq 0$. This implies, by the ergodicity of the Boolean model, that the event has probability either 0 or 1. As a result, we conclude that N is an a.s. constant (which can be infinity). $\square$

Note that this result immediately implies that unbounded components exist a.s. in the supercritical regime, because $P(N \geq 1)$ is positive and there must be some $1 \leq k \leq \infty$ for which $P(N = k) = 1$.

2.2 Coupling and scaling

In this section we introduce two important concepts in the theory of continuum percolation. As we shall see, *coupling* and *scaling* are strongly related to each other.

It is not so easy to give a concise and clear definition of coupling. Maybe one could say that coupling is the construction of different models on the same probability space in some sensible way, in order to compare the two models directly. This 'definition' is somewhat vague, and this is why we think that the best way of introducing coupling is by means of some examples. Let us first give an example which has nothing to do with percolation theory, but which is quite instructive. Suppose there are two players, inevitably called A and B, who each have a coin. The coin of player A has probability p_A of heads coming up; for player B, this probability is p_B. Suppose now that $p_A \geq p_B$, and suppose that we are interested in the expected number of tosses needed by either of the two players to see heads five times. It is intuitively clear that the expected number for A is no larger than the expected number for B. It is not hard to prove this by doing the right calculations, but we give a more elegant method based on coupling. Let U_i, for $i = 1, 2, \ldots$, be an i.i.d. sequence of uniform-$(0, 1)$ distributed random variables. We model the experiments above by saying that heads comes up for player A at the i-th toss if and only if $U_i \leq p_A$. For B the requirement is that $U_i \leq p_B$. In this way, the tosses of the players are coupled and no longer independent. This, however, does not affect the expected number we are interested in. It is clear from the construction that player A sees heads five times no later than B *surely*, and hence the expected number of tosses to see heads five times for A cannot be larger than the corresponding quantity for B. The coupling described here enables us to compare different players directly.

After this warm-up, we give a percolation application of coupling. We prove the claim made in Chapter 1 that the percolation function θ in the Poisson Boolean model (X, ρ, λ) is non-decreasing in λ, the density of the process.

Proposition 2.9 *If $\lambda_1 \leq \lambda_2$ and $\rho_1 \leq \rho_2$ a.s. then $\theta_{\rho_1}(\lambda_1) \leq \theta_{\rho_2}(\lambda_2)$.*

Proof To begin with, let $\rho_1 = \rho_2 = \rho$ a.s. and consider a Boolean model (X, ρ, λ_2). We thin this process as described in Proposition 1.3: each point of X is taken away (together with its associated ball) with probability $1 - \lambda_1(\lambda_2)^{-1}$. It follows from Proposition 1.3 that the ensuing point process is again Poisson with density λ_1. We remove all balls centred at the deleted points and leave the other points and balls unchanged. It is easy to see that the ensuing model is a Poisson Boolean model (X, ρ, λ_1). However, it is clear from the construction that we have coupled the two processes in such a way that the occupied region in (X, ρ, λ_1) is a subset of the occupied region in (X, ρ, λ_2). Hence the existence of an unbounded occupied component in (X, ρ, λ_1) implies the existence of an unbounded occupied component in (X, ρ, λ_2). The case when ρ_1 and ρ_2 are different is treated similarly. □

Another example of coupling which we shall use frequently is described in Figure 2.1. Here we place a Poisson point process X with density λ. Centred at each point we place two balls, one of radius $\rho(\omega)$ and the other of radius $a\rho(\omega)$. Thus we define two Boolean models (X, ρ, λ) and $(X, a\rho, \lambda)$ on the same probability space and compare them.

Related to the concept of coupling is the concept of scaling. It is again not easy to give a precise definition, but we can say that scaling involves changing the unit of length in the model in order to compare two different percolation

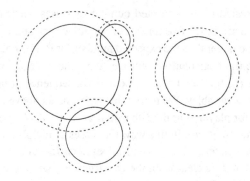

Figure 2.1. Coupling of the models (X, ρ, λ) (solid balls) and $(X, a\rho, \lambda)$ (dashed balls) for $a > 1$.

processes. It is often used in combination with coupling. An example works much better than any attempt to describe the concept, so here is a simple but important example.

Proposition 2.10 *Consider Poisson Boolean models in* $\mathbb{R}^d$. *Let, for* $r \geq 0$, $\lambda_c(r)$ *denote the critical density in the case where* $\rho = r$ *a.s. Then it is the case that*

$$\lambda_c(r_1)r_1^d = \lambda_c(r_2)r_2^d, \tag{2.13}$$

where $r_1, r_2 > 0$.

Proof   We discuss two approaches. First, consider the Boolean model (X, ρ_1, λ), where $\rho_1 = r_1$ a.s. This means that the expected number of points of the point process inside the unit cube is equal to λ. Now we rescale the model by doing the following: instead of looking at unit cubes, we tile the space with cubes of side length r_1/r_2. The volume of such a cube is $(r_1/r_2)^d$ and hence the expected number of points of X inside such a cube is $\lambda(r_1/r_2)^d$. Furthermore, the relative length of the radii compared to the side length of the new cubes is equal to $r_1(r_2/r_1) = r_2$. If we declare as our new unit of length the size of the new cubes $(= r_1(r_2)^{-1})$, then we see in fact a Boolean model with density $\lambda(r_1/r_2)^d$ and where the radii of the balls are equal to r_2 a.s.; i.e. we see the model $(X, \rho_2, \lambda(r_1/r_2)^d)$, where of course ρ_2 takes the value r_2 a.s. Note that we have not changed one single point or ball, we just look at the realisation from a different point of view. It follows that if (X, ρ_1, λ) is supercritical, then so is $(X, \rho_2, \lambda(r_1/r_2)^d)$, and if the former is subcritical, so is the latter. Hence, $\lambda_c(r_2) = \lambda_c(r_1)(r_1/r_2)^d$, and we are done.

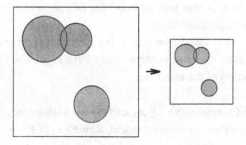

Figure 2.2. Scaling by a factor 2.

The second approach is to consider a realisation of the model (X, ρ_1, λ) and apply the transformation $x \to (r_2/r_1)x$ on the d-dimensional space. We then end up with a realisation of a Boolean model where all balls have radius r_2 and where the points form a Poisson process with density $\lambda(r_1/r_2)^d$ (see Figure 2.2). Note that the structure of the configuration is not affected by this transformation: in the new model, the same balls overlap as in the original model. This means that if the first model is subcritical, so is the second and vice versa. The conclusion now follows as before.   □

As follows from the proof of Proposition 2.10, the models (X, r_1, λ_1) and (X, r_2, λ_2) are related if $\lambda_2 = \lambda_1 (r_1/r_2)^d$ in the sense that the two models can be seen as scaled versions of each other. Any property in one of these models can be reformulated by a suitable scaling in the other model. This property is by no means restricted to Boolean models with fixed-size balls. The proof of the next proposition is similar to the proof of Proposition 2.10 and we omit it.

Proposition 2.11 *In any d-dimensional Poisson Boolean model we have, for any $a > 0$,*

$$\lambda_c^*(a\rho) = \lambda_c^*(\rho)a^{-d}$$

and

$$\lambda_c(a\rho) = \lambda_c(\rho)a^{-d}.$$

2.3 The FKG inequality

There is a natural definition of increasing and decreasing events in the present continuum setting. Consider two realisations ω and ω' of a Poisson Boolean model. We define a partial ordering '$\preceq$' as $\omega \preceq \omega'$ if and only if every ball

$S(x_i, \rho_i)$ present in ω is also present in ω'; i.e. ω' can be obtained from ω by adding points (and associated balls).

An event $A \in \Omega$ is said to be *increasing* (respectively *decreasing*) if for every $\omega \preceq \omega'$, $1_A(\omega) \le 1_A(\omega')$ (respectively $1_A(\omega) \ge 1_A(\omega')$), where 1_A is the indicator function of the event A.

Theorem 2.2 (FKG inequality)   *If A_1 and A_2 are both increasing or both decreasing events in a Poisson Boolean model, then $P(A_1 \cap A_2) \ge P(A_1)P(A_2)$.*

Proof   Consider the lattice $\mathbb{L}_n = (2^{-n}\mathbf{Z})^d \times (2^{-n}\mathbf{Z})$. For any $k = (k_1, \dots, k_d) \in \mathbf{Z}^d$ and $s \in \mathbf{Z}_+$ let $C(k, s)$ denote the cell $\{(x, r) \in \mathbb{R}^d \times \mathbb{R}_+ : (s-1)2^{-n} < r \le s2^{-n}$ and $(k_i - 1)2^{-n} < x_i \le k_i 2^{-n}$ for every $i = 1, \dots, d\}$, where $x = (x_1, \dots, x_d)$. Thus for distinct (k, s), the cells $C(k, s)$ are disjoint and also $\cup\, C(k, s) = \mathbb{R}^d \times \mathbb{R}_+$, where the union is over all $k \in \mathbf{Z}^d$ and $s \in \mathbf{Z}_+$. Given any cell $C = C(k, s)$, consider the random variable $N_n(C)$ which is defined as the number of Poisson points in $\prod_{i=1}^{d}((k_i - 1)2^{-n}, k_i 2^{-n}]$ whose associated ball has radius in $((s-1)2^{-n}, s2^{-n}]$. Let $\mathcal{F}_n$ be the σ-algebra generated by the random variables $\{N_n(C) : C$ is a cell in $\mathbb{L}_n\}$. Then for any event A, $\{E(1_A|\mathcal{F}_n), n \ge 1\}$ is a martingale with respect to $\mathcal{F}_n$ whence, by the martingale convergence theorem, we have

$$E(1_A|\mathcal{F}_n) \to 1_A \text{ a.s. as } n \to \infty. \tag{2.14}$$

For fixed n it is not hard to see that the random variables $\{N_n(C)\}$ are all independent. Also, $E(1_A|\mathcal{F}_n)(\omega)$ is a function of $\{N_n(C)\}_C$. As such, we easily see that if A is an increasing event, then $E(1_A|\mathcal{F}_n)$ is an increasing function. Thus, for any $\omega \preceq \omega'$, $E(1_A|\mathcal{F}_n)(\omega) \le E(1_A|\mathcal{F}_n)(\omega')$. Now for two increasing events A_1 and A_2, we have by the standard FKG inequality (Theorem 1.4),

$$E\{E(1_{A_1}|\mathcal{F}_n)E(1_{A_2}|\mathcal{F}_n)\} \ge E\{E(1_{A_1}|\mathcal{F}_n)\}E\{E(1_{A_2}|\mathcal{F}_n)\}$$

$$= E(1_{A_1})E(1_{A_2}). \tag{2.15}$$

Letting n tend to infinity and applying Lebesgue's dominated convergence theorem, we have from (2.14)

$$E\{E(1_{A_1}|\mathcal{F}_n)E(1_{A_2}|\mathcal{F}_n)\} \to E\{1_{A_1}1_{A_2}\}.$$

Thus from (2.15) we have $P(A_1 \cap A_2) \ge P(A_1)P(A_2)$. This completes the proof of the theorem. $\square$

As an application of the FKG inequality we give two examples.

Example 2.1   Let $B \subseteq I\!\!R^d$ be a bounded measurable set containing the origin. For $m \geq 0$ consider the following events in the Poisson Boolean model (X, ρ, λ):

$$E = \{d(W(B)) \geq m\}, \quad F = \{B \subseteq C\},$$

where, as defined in Chapter 1, $W(B)$ denotes the union of all occupied components which intersect B, $d(W(B))$ denotes the diameter $\sup\{d(x, y) : x, y \in W(B)\}$ and C the occupied region. Clearly, E and F are both increasing events, so by the FKG inequality we have $P_\lambda(E \cap F) \geq P_\lambda(E)P_\lambda(F)$. Thus

$$
\begin{aligned}
P_\lambda\{d(W) \geq m\} &\geq P_\lambda(E \cap F) \\
&\geq P_\lambda(E)P_\lambda(F) \\
&= C(\lambda, B)P_\lambda\{d(W(B)) \geq m\},
\end{aligned}
$$

where $C(\lambda, B) = P_\lambda(F) > 0$ because B is a bounded region. Thus we have the inequality

$$E_\lambda(d(W(B))) \leq K(\lambda, B)E_\lambda(d(W)), \tag{2.16}$$

for any bounded region B containing the origin and a positive constant $K(\lambda, B)$.

For the vacant case, a similar application of the FKG inequality yields

$$E_\lambda(d(V(B))) \leq K^*(\lambda, B)E_\lambda(d(V)), \tag{2.17}$$

for any bounded region B containing the origin and a positive constant $K^*(\lambda, B)$, where as in Chapter 1, $V(B)$ denotes the union of all vacant components which intersect B and V denotes the vacant component of the origin.

Example 2.2   We first define crossing probabilities of a box. Consider the d-dimensional box $B := [0, l_1] \times \cdots \times [0, l_d]$ and let $B_0(i) := [0, l_1] \times \cdots \times [0, l_{i-1}] \times \{0\} \times [0, l_{i+1}] \times \cdots \times [0, l_d]$ and $B_1(i) := [0, l_1] \times \cdots \times [0, l_{i-1}] \times \{l_i\} \times [0, l_{i+1}] \times \cdots \times [0, l_d]$ be two faces of the box B. For $1 \leq i \leq d$, we define the *occupied crossing probability in the i-th direction* as

$$\sigma((l_1, \ldots, l_d), \lambda, i) := P_\lambda\{\text{there is a continuous curve } \gamma \text{ in } B \text{ such that}$$

$$(i) \ \gamma \subseteq C \cap B$$

$$(ii) \ \gamma \cap B_0(i) \neq \emptyset \text{ and } \gamma \cap B_1(i) \neq \emptyset\}. \tag{2.18}$$

In two dimensions, for $i = 1$, the event in the parentheses in (2.18) is called a left–right (L–R) occupied crossing of the rectangle $[0, l_1] \times [0, l_2]$, and for $i = 2$ it is called a top–bottom (T–B) occupied crossing of the rectangle $[0, l_1] \times [0, l_2]$.

The vacant crossing probabilities are defined similarly. For $1 \leq i \leq d$, we define the *vacant crossing probability in the i-th direction* as

$$\sigma^*((l_1, \ldots, l_d), \lambda, i) := P_\lambda \{\text{there is a continuous curve } \gamma^* \text{ in } B \text{ such that}$$

$$(i) \quad \gamma^* \cap C = \emptyset$$

$$(ii) \quad \gamma^* \cap B_0(i) \neq \emptyset \text{ and } \gamma^* \cap B_1(i) \neq \emptyset\}. \quad (2.19)$$

In two dimensions, for $i = 1$, the event in the parentheses in (2.19) is called a left–right (L–R) vacant crossing of the rectangle $[0, l_1] \times [0, l_2]$, and for $i = 2$ it is called a top–bottom (T–B) vacant crossing of the rectangle $[0, l_1] \times [0, l_2]$.

By an application of the FKG inequality we have $P\{\text{there exists an L–R occupied and a T–B occupied crossing of } [0, l_1] \times [0, l_2]\} \geq P\{\text{there exists an L–R occupied crossing of } [0, l_1] \times [0, l_2]\} \, P\{\text{there exists a T–B occupied crossing of } [0, l_1] \times [0, l_2]\}$. A similar statement can be made for vacant crossings and in higher dimensions.

2.4 The BK inequality

In this section we shall prove a result which is the analogue of the BK inequality for discrete percolation. The inequality will in some sense be dual to the FKG inequality. In order to state the result we need some definitions. Any $\omega \in \Omega$ corresponds to a countable set of pairs $S(\omega) = \{(x_i, r_i)\}$ where the x_i's denote the points of the point process and r_i is the radius of the ball centred at x_i. For any bounded Borel set $U \subset \mathbb{R}^d$, we define

$$\omega_U = \{(x_i, r_i) : (x_i, r_i) \in S(\omega), x_i \in U\}.$$

The event $[\omega_U]$ is defined as

$$[\omega_U] = \{\omega' : \text{there exists } \omega'' \preceq \omega' \text{ such that } \omega''_U = \omega_U\}.$$

In words, this is the event that the configuration inside U is larger than ω_U. We say that an increasing event A is an *event on U* if $\omega \in A$ and $\omega' \in [\omega_U]$ imply that $\omega' \in A$. A *rational rectangle* is an open d-dimensional cube with rational coordinates.

Definition 2.4 *Let A and B be two increasing events on a bounded Borel set U. Then*

$$A \,\square\, B = \{\omega : \text{there are disjoint sets } V \text{ and } W \text{ such that } V$$

$$\text{and } W \text{ are finite unions of rational rectangles}$$

$$\text{and } [\omega_V] \subset A, [\omega_W] \subset B\}.$$

When $A \,\square\, B$ occurs, we say that A and B occur disjointly.

The fact that we only consider unions of rational rectangles in the definition has to do to with measurability problems but is otherwise unimportant. If a set V satisfies $[\omega_V] \subset A$, then any set containing V has the same property, and it is easy to see that there is also a subset of V with the same property. Hence no 'minimal set' with the required properties exists. Note also that this definition of $A \,\square\, B$ is equivalent to requiring the existence of two disjoint sets of points of the point process (in U) such that any configuration which agrees with ω on the first set of points (including the associated balls) is in A, and any configuration which agrees with ω on the second set of points is in B. Before proceeding we clarify the definition with an example.

Example 2.3   Consider a Boolean model in two dimensions and suppose that the radii are bounded from above by $R > 0$. Consider a rectangle $[0, l_1] \times [0, l_2]$. Let A be the event that there is an L–R occupied crossing of the rectangle, and let B be the event that there is an occupied T–B crossing. Then A and B are increasing events on $[-R, l_1 + R] \times [-R, l_2 + R]$, and $A \,\square\, B$ is the event that there exist both an L–R crossing and a T–B crossing in such a way that the balls used for the L–R crossing are different than the balls used for the T–B crossing. Note that the balls in one crossing are allowed to (and in this case certainly will) intersect balls in the other crossing.

Theorem 2.3 (BK inequality)   *Suppose U is a bounded measurable set in $\mathbb{R}^d$. For any two increasing events A and B on U in a Poisson Boolean model we have*

$$P(A \,\square\, B) \leq P(A)P(B).$$

The key to the proof of this theorem is an 'exchanging technique'. To explain this, consider two sets U and $x + U$ in $\mathbb{R}^d$, where U is bounded and x is chosen such that the two sets are disjoint. What we want to do is, given a realisation of the Boolean model, exchange the configurations on U and $x + U$ in the sense that all balls centred in U are moved to $x + U$ and vice versa. For our purposes it suffices to restrict ourselves to points x with integer coordinates. For such x we define, for $\omega \in \Omega$, $T_x^U(\omega)$ to be the configuration obtained from ω by (i) translating all points of the point process in U by x and all points in $x + U$ by $-x$, and (ii) for all points y of the point process in U, we interchange the values of the random variables corresponding to $K(n_0(y), z(n_0, y))$ and $K(n_0(y), z(n_0, y)) + x$, and for all points of the process in $U + x$ we interchange the values of the random variables corresponding to $K(n_0(y), z(n_0, y))$ and $K(n_0(y), z(n_0, y)) - x$. (Recall that $K(n_0(y), z(n_0, y))$ is the binary cube whose associated random variable gives the radius of the ball centered at y.) In the realisation corresponding to ω, this comes down to exchanging points and balls in U and $x + U$, as anticipated previously. For any event A, $T_x^U(A)$ is defined as $\{T_x^U(\omega) : \omega \in A\}$. In words, $T_x^U(A)$ is the event that A *would* occur if we were to interchange the realisations on U and $x + U$.

The following lemma relates $P(A \,\square\, B)$ to $P(A \,\square\, T_x^U(B))$. The second event can be described as the event that A occurs, and if we were to interchange the realisations on U and $x + U$, B would occur. Thus the event A can 'use' U, and the event B can 'use' $T_x^U(U)$. This fact should make it easier for A and B to occur disjointly, and the next lemma is the first step in this direction.

Lemma 2.2   *Let M be a bounded Borel set, $U \subset M$ and x a vector with integer coordinates such that $M \cap (x + U) = \emptyset$. For increasing events A and B on M we have*

$$P(A \,\square\, B) \le P(A \,\square\, T_x^U(B)) + P(X(U) \ge 2).$$

Proof   We define, for all bounded Borel sets U, the set $\Omega_U = \{\omega_U : \omega \in \Omega\}$. For any subset $\Gamma \subset \Omega_U$, we define $[\Gamma] = \cup_{\{\gamma \in \Gamma\}}[\gamma]$. For disjoint Borel sets U and U', we write $(\omega_U, \omega'_{U'})$ for the element in $\Omega_{U \cup U'}$ which agrees with ω on U and with ω' on U'. Also, for $\gamma \in \Omega_M$ and $V \subset M$ we write $[\gamma]_V$ to mean $[\gamma'_V]$ for any $\gamma' \in \Omega$ with $\gamma'_M = \gamma$. For $\alpha \in \Omega_{M \setminus U}$ let $A(\alpha)$ be the set $\{\sigma \in \Omega_U$: there exist $V, W \subset M, V \cap W = \emptyset, W \subset M \setminus U$ such that $[(\alpha, \sigma)]_V \subset A$ and $[(\alpha, \sigma)]_W \subset B\}$. (Here we assume again that V and W are finite unions of rational rectangles.) The event $B(\alpha)$ is defined similarly with the roles of A and B interchanged. In words, $A(\alpha)$ is the set of configurations in Ω_U which make $A \,\square\, B$ to occur in such a way that the set W corresponding to B is outside U. The event $B(\alpha)$ can be described similarly.

Now let $\omega \in A \,\square\, B$. If it is impossible to choose V and W in such a way that they have non-empty intersection with U, then it follows, using the fact that both A and B are increasing events, that $X(U) \ge 2$. Hence either $\omega_U \in A(\omega_{M \setminus U})$, $\omega_U \in B(\omega_{M \setminus U})$ or $X(U)(\omega) \ge 2$. Writing P' for the conditional probability measure given $\omega_{M \setminus U}$ we thus find a.s.

$$
\begin{aligned}
P'(A \,\square\, B) &\le P'([A(\omega_{M \setminus U})]) + P'([B(\omega_{M \setminus U})]) \\
&\quad - P'([A(\omega_{M \setminus U})] \cap [B(\omega_{M \setminus U})]) + P'(X(U) \ge 2) \\
&\le P'([A(\omega_{M \setminus U})]) + P'([B(\omega_{M \setminus U})]) \\
&\quad - P'([A(\omega_{M \setminus U})])P'([B(\omega_{M \setminus U})]) + P'(X(U) \ge 2), \quad (2.20)
\end{aligned}
$$

where the last inequality follows from the FKG inequality. Note that we can apply the FKG inequality because the events involved are increasing and not affected by the conditioning.

Furthermore, if $\omega_U \in A(\omega_{M \setminus U})$ then, according to the definitions, $\omega \in A \square B$ in such a way that the set W corresponding to B can be taken outside U. Hence after interchanging the configurations on U and $x + U$, B still occurs whence $\omega \in A \square T_x^U(B)$.

If $\omega_{x+U} \in T_x^U(B(\omega_{M \setminus U}))$ then after interchanging U and $x + U$, $B(\omega_{M \setminus U})$ occurs, which means that after interchanging, $A \square B$ occurs in such a way that the set V corresponding to A is outside U. This implies that $\omega \in A \square T_x^U(B)$.

From the last two paragraphs it follows that a.s.

$$
\begin{aligned}
P'(A \square T_x^U(B)) &\geq P'([A(\omega_{M \setminus U})]) + P'(T_x^U([B(\omega_{M \setminus U})])) \\
&\quad - P'([A(\omega_{M \setminus U})] \cap T_x^U([B(\omega_{M \setminus U})])) \\
&= P'([A(\omega_{M \setminus U})]) + P'(T_x^U([B(\omega_{M \setminus U})])) \\
&\quad - P'([A(\omega_{M \setminus U})]) P'(T_x^U([B(\omega_{M \setminus U})])) \\
&= P'([A(\omega_{M \setminus U})]) + P'([B(\omega_{M \setminus U})]) \\
&\quad - P'([A(\omega_{M \setminus U})]) P'(T_x^U([B(\omega_{M \setminus U})])), \quad (2.21)
\end{aligned}
$$

where the first equality follows from independence and the second from stationarity. From (2.20) and (2.21) we find

$$ P'(A \square B) \leq P'(A \square T_x^U(B)) + P'(X(U) \geq 2) $$

from which the lemma follows after integrating out the conditioning on $\omega_{M \setminus U}$. $\square$

Proof of Theorem 2.3 Using a simple scaling argument, we can assume without loss of generality that A and B are increasing events on the unit cube $I = [0, 1]^d$. Choose any x with integer coordinates such that $(x + I) \cap I = \emptyset$. Fix an integer n and partition I into 2nd binary cubes $S_1, \ldots, S_{2^{nd}}$. Define the events $B^{(k)}$ for $k = 1, \ldots, 2^{nd}$ as follows: $B^{(0)} = B$, $B^{(k+1)} = T_x^{S_{k+1}}(B^{(k)})$, for $k = 0, \ldots, 2^{nd} - 1$. Note that with this definition, $B^{(2^{nd})} = T_x^I(B)$. Applying Lemma 2.2 2nd times (the k-th time with B replaced by $B^{(k-1)}$, $M = U \cup \{x + (S_1 \cup \cdots \cup S_{k-1})\}$ and $U = S_k$), we obtain

$$
\begin{aligned}
P(A \square B) = P(A \square B^{(0)}) &\leq P(A \square B^{(1)}) + P(X(S_1) \geq 2) \\
&\leq \cdots \leq P(A \square B^{(2^{nd})}) + 2^{nd} P(X(S_1) \geq 2) \\
&\leq P(A)P(B) + 2^{nd}(\lambda 2^{-nd})^2 \\
&= P(A)P(B) + \lambda^2 2^{-nd}.
\end{aligned}
$$

As n is arbitrary, we now take the limit for $n \to \infty$ and the proof is complete. $\square$

We give a typical application of the BK inequality which we shall need later on.

Theorem 2.4 *Consider a Poisson Boolean model (X, ρ, λ) where ρ satisfies $0 \leq \rho \leq R$ for some $R < \infty$. If $E(d(W)) < \infty$ then there exist constants $C_1, C_2 > 0$, depending on λ and the dimension d such that*

$$P_\lambda(0 \overset{o}{\rightsquigarrow} (B_m)^c) \leq C_1 e^{-C_2 m}.$$

Proof It follows from Example 2.1 that $E(d(W(B))) < \infty$ for any bounded set B and it will be enough to show that under this condition $P(B \overset{o}{\rightsquigarrow} (B_m)^c) \leq C_1 e^{-C_2 m}$. We partition the space with cubes of the form

$$B_R(z) = \prod_{i=1}^{d} (2Rz_i - R, 2Rz_i + R],$$

where $z \in \mathbf{Z}^d$. Since $E(d(W(B_R(0)))) < \infty$ we can choose M so large that $E(W_M) < \frac{1}{2} 3^{-d}$, where W_M is the number of cubes $B_R(z)$ outside B_M which intersect $W(B_R(0))$. Now choose L so large that the set $\cup_{||z|| \geq L-1} B_R(z)$ is disjoint from B_M, where $||z|| = \max_i z_i$. Then choose m so large that $B_m \supset \cup_{||z|| \leq L+1} B_R(z)$. Observe that if $\{B_R(0) \overset{o}{\rightsquigarrow} (B_m)^c\}$ occurs, then there is some z with $||z|| = L$ for which $\{B_R(0) \overset{o}{\rightsquigarrow} D(z)\}$ and $\{B_R(z) \overset{o}{\rightsquigarrow} (B_m)^c\}$ occur disjointly, where $D(z)$ is defined to be the set $D(z) = \cup_{||z-z'||=1} B_R(z')$. It then follows from the BK inequality that

$$P(B_R(0) \overset{o}{\rightsquigarrow} (B_m)^c) \leq \sum_{\{z:||z||=L\}} P(B_R(0) \overset{o}{\rightsquigarrow} D(z), B_R(z) \overset{o}{\rightsquigarrow} (B_m)^c)$$

$$\leq \max_{\{z:||z||=L\}} P(B_R(z) \overset{o}{\rightsquigarrow} (B_m)^c) \times$$

$$\times \sum_{\{z:||z||=L\}} P(B_R(0) \overset{o}{\rightsquigarrow} D(z))$$

$$\leq \max_{\{z:||z||=L\}} P(B_R(z) \overset{o}{\rightsquigarrow} (B_m)^c) 3^d E(W_M),$$

where the last inequality follows from the fact that any cube is contained in at most 3^d sets $D(z)$ with $||z|| = L$. It follows that

$$P(B_R(0) \overset{o}{\rightsquigarrow} (B_m)^c) \leq \frac{1}{2} \max_{\{z:||z||=L\}} P(B_R(z) \overset{o}{\rightsquigarrow} (B_m)^c). \tag{2.22}$$

To estimate the right-hand side of (2.22), suppose that m is so large that $\cup_{\{z':||z-z'||=L\}} \subset B_m$. Then the same argument as above shows that for all

z with $||z|| = L$,

$$P(B_R(z) \overset{o}{\leadsto} (B_m)^c) \leq \frac{1}{2} \max_{\{z':||z-z'||=L\}} P(B_R(z') \overset{o}{\leadsto} (B_m)^c).$$

Repeating this argument now leads to the desired conclusion. □

2.5 Notes

Proposition 2.7 is from Pugh and Shub (1971). The FKG inequality for continuum percolation appears in Roy (1988). The proof of the BK inequality given here is due to v.d. Berg (1995), and is in fact a proof of the BK inequality for marked point processes. In Roy and Sarkar (1993), a more restricted version of the BK inequality is proved for certain classes of increasing events. Also in Bezuidenhout and Grimmett (1991), a version of the continuum BK inequality appears. Coupling methods are very old and have become quite popular in recent years. The scaling relations in Proposition 2.11 appear in Zuev and Sidorenko (1985).

3

Occupancy in Poisson Boolean models

The percolation-theoretical study of the Boolean model confines itself primarily to the study of the geometric and probabilistic properties of the occupied and vacant clusters. In this chapter we shall study the occupied region in a Poisson Boolean model (X, ρ, λ). Throughout this chapter (X, ρ, λ) will denote a Poisson Boolean model arising from an underlying Poisson point process X of density λ and radius random variable ρ. As usual we shall assume that centred at points $x_1, x_2, \ldots$ of X are spheres $S(x_1), S(x_2), \ldots$ of radius $\rho_1, \rho_2, \ldots$, respectively, where $\rho_1, \rho_2, \ldots$ are independent and identically distributed and are also independent of the underlying point process X. Let ρ denote a random variable independent of $\rho_1, \rho_2, \ldots$ and also independent of the process X and whose distribution is identical to that of ρ_1. With a slight abuse of notation we shall let P_λ denote the probability measure governing this Poisson Boolean model.

3.1 Introduction

It is quite possible that the Boolean model is such that the space $I\!\!R^d$ is completely covered. To this end, we first give the following elementary result:

Lemma 3.1 *Suppose that $E\rho^d < \infty$. Then the number of balls which intersect $S(0, t)$ (the ball with radius t centred at the origin) has a Poisson distribution with finite parameter*

$$\lambda \int_{I\!\!R^d} P(\rho \geq |x| - t)dx. \tag{3.1}$$

Proof It follows immediately from Proposition 1.3 and the independence of the radii and the point process that the random variable in question has a

Poisson distribution with parameter given by (3.1). We need only show that this parameter is finite. For this, we have, writing F_ρ for the distribution function of the radius random variable ρ, and for some positive constants C and C':

$$\int_{\mathbb{R}^d} P(\rho \geq |x| - t)dx = C \int_0^\infty r^{d-1} P(\rho \geq r - t)dr$$

$$= C \int_0^\infty r^{d-1} \int_{r-t}^\infty dF_\rho(y)dr$$

$$= C \int_0^\infty \int_0^{y+t} r^{d-1} dr dF_\rho(y)$$

$$= C' \int_0^\infty (y + t)^d dF_\rho(y) < \infty. \qquad \square$$

Taking $t = 0$, we see that the expected number of balls which intersect the origin is equal to $\lambda \int_{\mathbb{R}^d} P(\rho \geq |x|)dx$. The following argument shows that this is equal to the expected number of Poisson points in the random ball around the origin (assuming that the point process is conditioned to have a point at the origin):

$$\lambda \int_{\mathbb{R}^d} P(\rho \geq |x|)dx = E\left(\sum_i 1_{\{S(x_i, \rho_i) \cap \{0\} \neq \emptyset\}}\right)$$

$$= E\left(\sum_i 1_{\{x_i \in S(0, \rho_i)\}}\right)$$

$$= E\left(\sum_i 1_{\{x_i \in S(0, \rho)\}}\right)$$

$$= \lambda \pi_d E\rho^d, \qquad (3.2)$$

where π_d denotes the volume of a d-dimensional ball with unit radius, and where the last equality follows from conditioning on ρ. We conclude that the probability that the origin is not covered is equal to

$$P(W = \emptyset) = e^{-\lambda \pi_d E\rho^d}. \qquad (3.3)$$

The question of complete coverage is settled in the next proposition.

Proposition 3.1 *In a Poisson Boolean model (X, ρ, λ) on $\mathbb{R}^d$, the whole space is covered a.s. if and only if $E\rho^d = \infty$.*

Proof Let us denote the vacant region inside the unit cube $[0, 1]^d$ by V'. We then have, by (3.3) and Fubini's theorem,

$$E(\ell(V')) = E \int_{[0,1]^d} 1_{\{x \text{ is not covered}\}} \, dx$$
$$= e^{-\lambda \pi_d E \rho^d}.$$

Hence if $E\rho^d < \infty$, then $E(\ell(V')) > 0$ and thus $P(\ell(V') > 0) > 0$. Using Proposition 2.8 and Proposition 2.2 with T_i the translation over the i-th unit vector and $f(\omega) = \ell(V')(\omega)$, we see that the space is almost surely not completely covered.

Conversely, if $E\rho^d = \infty$, then the vacancy in the unit cube has expected Lebesgue measure zero. We distinguish between two possible cases.

Suppose first that any bounded region is intersected by only finitely many balls a.s. We then cover the space with countably many overlapping open sets of the form $\frac{1}{2}z + (0, 1)^d =: D(z)$, for $z \in \mathbf{Z}^d$. The vacancy inside $D(z)$ is an open set which has measure zero a.s. There is only one such open set, the empty set, and we conclude that the whole space is covered a.s.

Next suppose that there is a $t > 0$ such that $S(0, t)$ is intersected by infinitely many balls with positive probability. If this happens then there exists a half-line l starting at the origin and a sequence $(x_1, x_2, \ldots)$ of points of the point process X such that each of the balls $S(x_i)$ intersects $S(0, t)$, and such that the angle between l and the line passing through the origin and x_i tends to zero as i tends to infinity. However, since we must have $|x_i| \to \infty$ as $i \to \infty$, this implies that there is a half-space which is completely covered by balls. By rotation invariance, all transformations S_{e_i} act ergodically. (The notation is as in Chapter 2.) Consider random variables Y_n and Z_n, $n \in Z$ defined as follows: if $[0, 1]^{d-1} \times [n, n+1]$ is completely covered by balls, then $Y_n = 1$, otherwise $Y_n = 0$; if $[n, n+1] \times [0, 1]^{d-1}$ is completely covered then $Z_n = 1$, otherwise $Z_n = 0$. If a half-space is completely covered by balls, then one of the following possibilities occurs:

(i) $Y_n = 1$ for all n large enough or $Y_n = 1$ for all $-n$ large enough,

(ii) $Z_n = 1$ for all n large enough or $Z_n = 1$ for all $-n$ large enough.

Suppose (i) occurs. If $f := 1_{\{[0,1]^d \text{ is completely covered}\}}$ it follows from the ergodicity of S_{e_d} and Proposition 2.1 that $Ef = 1$, which implies that the unit cube is completely covered almost surely, which in turn implies by stationarity that the whole space is covered by balls a.s. If (ii) occurs the result follows similarly.   □

In view of this proposition we restrict our study to those random variables ρ for which

$$E\rho^d < \infty. \tag{3.4}$$

Suppose (3.4) holds. We define a stochastic process $\{X_n\}$ as follows:

$$X_n := \begin{cases} 1 & \text{if the point } (n, 0, \ldots, 0) \text{ is not covered in } (X, \rho, \lambda), \\ 0 & \text{otherwise.} \end{cases}$$

The translation by the vector e_1 is ergodic, and (3.3) together with Proposition 2.1 gives that $P(X_n = 1 \text{ i.o.}) = 1$. This yields

Corollary 3.1 *If (3.4) holds for a Poisson Boolean model (X, ρ, λ) on $\mathbb{R}^d$, then, for any $n \geq 1$, $P_\lambda((B_n)^c \subseteq W) = 0$.*

3.2 One-dimensional triviality

It is quite easy to provide a complete description of the percolation phenomenon in one-dimensional Boolean models. Indeed, from Corollary 3.1 we trivially have:

Theorem 3.1 *For a Poisson Boolean model (X, ρ, λ) on $\mathbb{R}$, with ρ satisfying (3.4) for $d = 1$, unbounded components do not occur a.s.*

Thus for a one-dimensional Poisson Boolean model, irrespective of the density of the underlying driving process, either there is complete coverage or no unbounded component, depending on the distribution of the radius random variable.

At this stage of the development of this chapter, we point out the importance of the geometric structure of the random shape at each point of the driving process. Let

$$S_n = \bigcup_{i=-n^2}^{n^2} \left[\frac{i}{n} - (2n^2 + 1)^{-1}, \frac{i}{n} + (2n^2 + 1)^{-1} \right]$$

for $n \geq 1$. Let $x_1, x_2, \ldots$ be an enumeration of the points of a Poisson point process on $\mathbb{R}$ and let $S(x_i) = x_i + S_i$ be the shape centred at the point x_i, where the distribution of S_i is given by

$$P(S_i = S_n) = \frac{6}{\pi} n^{-2} \quad \text{for } n \geq 1.$$

We also assume that $S_1, S_2, \ldots$ are i.i.d. and independent of the driving process. Observe that $\ell(S_n) = 2$ for every $n \geq 1$, so that $\ell(S(x_i)) = 2$ for every $i \geq 1$, and hence

$$E\ell(S(x_i)) = 2 \quad \text{for all } x_i \in X. \tag{3.5}$$

Note the condition (3.4) for $d = 1$ for Poisson Boolean models on $\mathbb{R}$ is similar to (3.5) in the sense that both guarantee that the expected Lebesgue measure of the associated shape $S(x_i)$ is finite. We will show that each random shape S_i is intersected by infinitely many other random shapes almost surely.

Consider an interval $(-a, a)$ for $a > 0$. Let $N_1, N_2, \ldots$ be i.i.d. random variables defined by $N_i = k$ if and only if $S_i = x_i + S_k$ for every $i \geq 1$ and $k \geq 1$. We note that whenever $|x_i| \leq k$ and $k \geq \lceil 1/2a \rceil + 1$ (where $\lceil x \rceil$ denotes the smallest integer greater than or equal to x), the shape $x_i + S_k$ has non-empty intersection with the interval $(-a, a)$. Hence, if

$$P_\lambda(|x_i| \leq N_i \text{ i.o.}) = 1, \tag{3.6}$$

then, with probability 1, infinitely many shapes have non-empty intersection with $(-a, a)$. This would, of course, prove our contention that each random shape is intersected by infinitely many random shapes with probability 1. To prove (3.6), we define, for every integer j,

$$E_j := \{x_i \in (j - 1, j) \text{ for some } i \geq 1 \text{ and } j \leq N_i\}.$$

Now,

$$P_\lambda(E_j) \geq P_\lambda\{x_i \in (j - 1, j) \text{ for some } i \geq 1\} P_\lambda(j \leq N_i)$$
$$= (1 - \exp(-\lambda)) \frac{6}{\pi} \sum_{n=j}^{\infty} n^{-2},$$

so $\sum_{j=1}^{\infty} P_\lambda(E_j) = \infty$. Thus by the Borel–Cantelli lemma

$$P_\lambda(E_j \text{ i.o. }) = 1.$$

However,

$$P_\lambda(|x_i| \leq N_i \text{ i.o.}) \geq P_\lambda(E_j \text{ i.o.}),$$

which shows that (3.6) holds.

This example shows that if we allow different shapes then the natural analogue of Theorem 3.1 need not hold.

3.3 Critical phenomena

In two and higher dimensions, we cannot provide a complete description as in Theorem 3.1.

Theorem 3.2 Let (X, ρ, λ) be a Poisson Boolean model on $\mathbb{R}^d$, for $d \geq 2$, with ρ satisfying (3.4). There exists $\lambda_0 > 0$ such that the expected number of balls in the component W which contains the origin is finite whenever $0 < \lambda < \lambda_0$ if and only if

$$E\rho^{2d} < \infty. \tag{3.7}$$

Proof Suppose $E\rho^{2d} < \infty$ and first assume that ρ takes only non-negative integer values. We shall employ an argument based on a comparison with a suitable branching process model to provide an upper bound on the expected number of balls contained in a component. The branching process we construct below is of multi-type (see Athreya and Ney 1972, chap. V, for the necessary theory). Suppose there is a ball S of radius i centred at x. Let n_j denote the (random) number of balls of radius j which intersect S. Since the Boolean model (X, ρ, λ) can be thought of being the superposition of the independent Boolean models $\{(X_j, j, \lambda P(\rho = j)); j = 0, 1, 2, \ldots\}$ we have that $n_1, n_2, \ldots$ is an independent sequence of random variables with each n_j having a Poisson distribution with mean

$$\mu_{i,j} := E_\lambda(\text{number of balls of radius } j \text{ centred in}$$
$$\{z : |z - x| \leq i + j\})$$
$$= \lambda P(\rho = j)\pi_d(i + j)^d, \tag{3.8}$$

where π_d is the d-dimensional volume of a unit ball. In our branching process n_j will represent the number of children of x of type j. More specifically, the member of the 0-th generation of the branching process is taken to be the origin. Here we place a ball S of random radius ρ_0, where ρ_0 and ρ are independent and have the same distribution. Now consider independent Poisson processes $X_{1,0}, X_{1,1}, X_{1,2}, \ldots$ with $X_{i,j}$ of density $\lambda P(\rho = j)$, $j = 0, 1, 2, \ldots$. Let $\{x_{1,j,k}\}$, $k = 0, 1, 2, \ldots, n_j$, be all the points of $X_{1,j}$ such that a ball of radius j centred at $x_{1,j,k}$ has non-empty intersection with S. The points $\{x_{1,j,k}; k = 0, 1, 2, \ldots, n_j, j = 0, 1, 2, \ldots\}$ are taken to be members of the first generation, where there are n_j members $\{x_{1,j,k}; k = 0, 1, 2, \ldots, n_j\}$ of type j. Since the superposition of the processes $\{X_{1,j}; j \geq 1\}$ yields a Boolean model with radius random variable ρ, Lemma 3.1 tells us that the total number of balls of all these processes which intersect a given bounded region is finite almost surely. Therefore, $\sum_{j=0}^{\infty} n_j$ is finite a.s. Also, as argued earlier, for all $j = 0, 1, 2, \ldots,$

$$E_\lambda(n_j | \rho_0 = i) = \lambda P(\rho = j) \pi_d (i + j)^d$$
$$= \mu_{i,j}.$$

Inductively, let $x_1, x_2, \ldots, x_m$ be the members of the n-th generation. Consider the member x_l, and suppose it is of type i. The members of the $(n + 1)$-th generation which are children of x_l are obtained by placing independent Poisson processes $X_0, X_1, X_2, \ldots$ of densities $\lambda P(\rho = 0), \lambda P(\rho = 1), \lambda P(\rho = 2), \ldots$ respectively and these Poisson processes are also independent of all other random processes already constructed. The children of x_l of type j are all those points of the process X_j such that a ball of radius j centred at any of these points will have non-empty intersection with a ball of radius i placed at x_l, where i was assumed to be the type of x_l. Thus given that x_l is of type i, the expected total number of children of x_l of type j, for $j = 0, 1, 2, \ldots$ is $\mu_{i,j}$.

Let $v_j^{(n)}$ be the expected number of members of the n-th generation of type j and let $v^{(n)}$ denote the infinite vector whose j-th entry is $v_j^{(n)}$. Also let M be the matrix with an infinite number of rows and an infinite number of columns, whose (i, j)-th entry is $\mu_{i,j}$. By the theory of multi-type branching processes (see sec. 1.6 of Mode 1971) conditioned on the member of the 0-th generation being of type i, the evolution of the process is given by

$$v^{(n)} = \mathbf{i} M^n, \tag{3.9}$$

where $\mathbf{i}$ denotes the infinite unit row vector whose i-th entry is 1 and all other entries are 0.

Let $\mu_{i,j}^{(n)}$ denote the (i, j)-th entry of the matrix M^n. Conditioned on the 0-th generation member being of type i, from (3.9) we have that the expected number of members of the n-th generation is

$$\sum_{j=0}^{\infty} v_j^{(n)} = \sum_{j=0}^{\infty} \mu_{i,j}^{(n)},$$

and the total expected number of members, μ_i, in the entire branching process is

$$\mu_i = \sum_{n=0}^{\infty} \sum_{j=0}^{\infty} v_j^{(n)} = \sum_{n=0}^{\infty} \sum_{j=0}^{\infty} \mu_{i,j}^{(n)}.$$

Now, for all $i, j \geq 1$, $(i + j)^d \leq 2^d [\max(i, j)]^d \leq 2^d i^d j^d$; thus, taking $C = 2^d \pi_d$,

$$\mu_{i,j} \leq \begin{cases} C\lambda P(\rho = j) i^d j^d & \text{for } i, j \geq 1 \\ C\lambda P(\rho = j) j^d & \text{for } i = 0 \\ C\lambda P(\rho = 0) i^d & \text{for } j = 0. \end{cases}$$

Hence, the (i, j)-th entry $\mu_{i,j}^{(2)}$ of M^2 satisfies, for $i, j \geq 1$,

$$\begin{aligned} \mu_{i,j}^{(2)} &= \sum_{l=0}^{\infty} \mu_{i,l} \mu_{l,j} \\ &\leq \sum_{l=0}^{\infty} C^2 \lambda^2 i^d j^d l^{2d} P(\rho = j) P(\rho = l) \\ &= C^2 \lambda^2 i^d j^d P(\rho = j) E\rho^{2d}, \end{aligned}$$

while, for $i = 0$,

$$\mu_{i,j}^{(2)} \leq C^2 \lambda^2 j^d P(\rho = j) E\rho^{2d}$$

and, for $j = 0$,

$$\mu_{i,j}^{(2)} \leq C^2 \lambda^2 i^d P(\rho = 0) E\rho^{2d}.$$

Inductively, it is easy to see that

$$\mu_{i,j}^{(n)} \leq \begin{cases} (C\lambda)^n (E\rho^{2d})^{n-1} i^d j^d P(\rho = j) & \text{for } i, j \geq 1 \\ (C\lambda)^n (E\rho^{2d})^{n-1} j^d P(\rho = j) & \text{for } i = 0 \\ (C\lambda)^n (E\rho^{2d})^{n-1} i^d P(\rho = 0) & \text{for } j = 0. \end{cases}$$

Since $E\rho^{2d} < \infty$, we have for $i \geq 1$,

$$\mu_i \leq 1 + i^d \sum_{n=1}^{\infty} (C\lambda)^n (E\rho^{2d})^{n-1} \left[P(\rho = 0) + \sum_{j=1}^{\infty} j^d P(\rho = j) \right] \quad (3.10)$$

and for $i = 0$,

$$\mu_i \leq 1 + \sum_{n=1}^{\infty} (C\lambda)^n (E\rho^{2d})^{n-1} \left[P(\rho = 0) + \sum_{j=1}^{\infty} j^d P(\rho = j) \right]. \quad (3.11)$$

Thus if $C\lambda E\rho^{2d} < 1$, then $\mu_i < \infty$.

Comparing this branching process to the Boolean model, the expected number of balls in the Boolean model which are in the component containing the ball at the origin is at most the total number of members in the branching process. Consequently, if $\lambda \leq (2^d \pi_d E\rho^{2d})^{-1}$, then the expected number of balls comprising a component is finite.

In case the radius random variable takes values other than integers, then we consider the Boolean model (X, ρ_{int}, λ), where $\rho_{int} = \lceil \rho \rceil$ denotes the smallest integer larger or equal than ρ. Clearly, $E\rho^{2d} < \infty$ implies $E\rho_{int}^{2d} < \infty$. By a coupling argument, it is immediate that the expected number of balls in a component of the model $(X\rho_{int}, \lambda)$ is at least that in a component of the model (X, ρ, λ), thereby proving the necessary part of the theorem.

To show the sufficiency part of the theorem, we assume that

$$E\rho^{2d} = \infty, \tag{3.12}$$

and prove that in the resulting Boolean model the expected number of balls in a component is infinite for every $\lambda > 0$. As before, we assume that the radius random variable ρ takes only non-negative integer values. The general case can be dispensed with by observing that if $\rho_{int} = \lfloor \rho \rfloor$ denotes the largest integer smaller than or equal to ρ, then $E\rho^{2d} = \infty$ implies that $E\rho_{int}^{2d} = \infty$, and the expected number of balls in a component of the model (X, ρ_{int}, λ) is at most that in a component of the model (X, ρ, λ).

Again, as before, let S be a ball of radius i centred at x and let n_j be the (random) number of balls of radius j in the Boolean model (X, ρ, λ) which intersect S. As in the argument leading to (3.8), we have that $n_1, n_2, \ldots$ is a sequence of independent random variables with each n_j having a Poisson distribution with mean given by (3.8). Let $M = \max\{j : n_j > 0\}$ and define $M = -1$ if $n_j = 0$ for all $j \geq 0$. Now

$$\{M = m\} = \{n_m > 0 \text{ and } n_j = 0 \text{ for all } j \geq m + 1\}, \tag{3.13}$$

so the event $\{M = m\}$ depends only on the Boolean models $\{(X_j, j, \lambda P(\rho = j)); j \geq m\}$. Thus the events $\{M = m\}$ and $\{M \geq m\}$ are independent of the Boolean models $\{(X_j, j, \lambda P(\rho = j)); j = 0, 1, 2, \ldots, m-1\}$. Let $k_0 = \min\{j \geq 1 : P(\rho = j) > 0\}$. Given S, a ball of radius i, and given $M \geq k_0 + 1$, let S' be a ball of radius M which has non-empty intersection with S. Then we have

$E_\lambda(\text{number of balls in the component containing } S)$

$$\geq \sum_{m=k_0+1}^{\infty} P(M = m)$$

$$\times E_\lambda(\text{number of balls of radius } k_0 \text{ intersecting } S' \mid M = m). \tag{3.14}$$

The last term can be bounded easily, observing that from the independence property described earlier, given $M = m$ and $m \geq k_0 + 1$, the number of balls of radius k_0 which intersect S' has a Poisson distribution with mean $\lambda P(\rho = k_0)\pi_d(m + k_0)^d$. Also,

$$
\begin{aligned}
P(M = m) &= P(n_m > 0)P(n_j = 0 \text{ for all } j \geq m + 1) \\
&= P(n_m > 0) \prod_{j \geq m+1} P(n_j = 0) \\
&\geq (1 - \exp(-\mu_{i,m})) \exp\left(-\sum_{j=m+1}^{\infty} \mu_{i,j}\right).
\end{aligned}
$$

Thus, from (3.14),

E_λ(number of balls in the component containing S)

$$
\begin{aligned}
&\geq \sum_{m=k_0+1}^{\infty} (1 - \exp(-\mu_{i,m})) \exp\left(-\sum_{j=m+1}^{\infty} \mu_{i,j}\right) \lambda P(\rho = k_0)\pi_d(m + k_0)^d \\
&\geq \sum_{m=k_0+1}^{\infty} \lambda \pi_d P(\rho = k_0)(m + k_0)^d (1 - \exp(-\mu_{i,m})) \\
&\quad \times \exp\left(-\sum_{j=0}^{\infty} \mu_{i,j}\right).
\end{aligned}
\tag{3.15}
$$

However, since $E\rho^d < \infty$,

$$
\sum_{j=0}^{\infty} \mu_{i,j} \leq C_1(\lambda) \sum_{j=0}^{\infty} (i + j)^d P(\rho = j) < \infty,
\tag{3.16}
$$

and, from (3.16)

$$
\begin{aligned}
1 - \exp(-\mu_{i,m}) &\geq \mu_{i,m} \exp(-\mu_{i,m}) \\
&\geq \lambda P(\rho = m)\pi_d m^d \exp\left(-\sum_{j=0}^{\infty} \mu_{i,j}\right) \\
&\geq C_2(\lambda)m^d P(\rho = m),
\end{aligned}
\tag{3.17}
$$

where $C_1(\lambda)$ and $C_2(\lambda)$ are positive constants.

In view of our assumption (3.12), combining the bounds obtained in (3.14), (3.15), (3.16) and (3.17), we have

$$E_\lambda(\text{number of balls in the component containing } S)$$

$$\geq C(\lambda) \sum_{m=k_0+1}^{\infty} m^{2d} P(\rho = m)$$

$$= \infty$$

where $C(\lambda)$ is a positive constant. This completes the proof of the theorem. $\square$

While (3.7) is necessary and sufficient for the finiteness of the expected number of balls in a component for sufficiently small densities, the following theorem asserts that (3.7) is not necessary for the component to be finite with probability 1.

Theorem 3.3 *For a Poisson Boolean model (X, ρ, λ) on $\mathbb{R}^d$, for $d \geq 2$, if $E\rho^{2d-1} < \infty$, then there exists $0 < \lambda_0$ such that for all $0 < \lambda < \lambda_0$, $P_\lambda(\text{number of balls in any occupied component is finite}) = 1$.*

Proof As in the previous theorem, we shall construct a multi-type branching process to estimate the number of balls in a component. However, we have to be more careful in the construction to enable us to obtain a better estimate. Also as before, it suffices to assume that ρ takes only positive integer values. Indeed, if $\rho_{int} = \lceil \rho \rceil$ denotes the smallest integer larger than or equal to ρ, then $E(\rho^{2d-1}) < \infty$ implies $E(\rho_{int}^{2d-1}) < \infty$.

In the new construction, the children of type j of an initial ball S of type i are all those balls of radius j which have non-empty intersection with S and which are not completely contained in S. The number n_j of such balls is clearly a Poisson random variable with mean

$$\mu_{i,j} := \lambda P(\rho = j)\pi_d[(i + j)^d - \{\max(0, i - j)\}^d], \tag{3.18}$$

and $n_1, n_2, \ldots$ are independent. Using this construction we will obtain an upper bound for the expected number of balls which make up the boundary of the component of S.

As in the proof of Theorem 3.2, we construct a branching process with this type distribution; the only difference is that the associated ball of a child cannot be completely covered by its immediate forebear. Thus we obtain the equation (3.9) for this branching process, where $v^{(n)} = (v_1^{(n)}, v_2^{(n)}, \ldots)$ and $M = ((\mu_{i,j}))$ with $v_j^{(n)}$ being the expected number of members in the n-th generation of type j in this branching process and $\mu_{i,j}$ is as defined in (3.18).

Now observe that, for $i \le j$,

$$(i + j)^d - \max\{0, (i - j)\}^d = (i + j)^d$$
$$\le 2^d j^d$$

while, for $i > j$,

$$(i + j)^d - \max\{0, (i - j)\}^d = \sum_{k=0}^{d} \binom{d}{k} i^{d-k} j^k - \sum_{k=0}^{d} (-1)^k \binom{d}{k} i^{d-k} j^k$$

$$\le \sum_{k=1}^{d} 2 \binom{d}{k} i^{d-k} j^k$$

$$\le 2^{d+1} i^{d-1} j^d,$$

where the last inequality holds because $\sum_{k=0}^{d} \binom{d}{k} = 2^d$.

Using the preceding inequalities, we have from (3.18), for every $i, j \ge 1$,

$$\mu_{i,j} \le C\lambda P(\rho = j) i^{d-1} j^d,$$

where C is a positive constant.

Performing a calculation as in the previous theorem, we see that

$$\mu_{i,j}^{(2)} = \sum_{l=0}^{\infty} \mu_{i,l} \mu_{l,j}$$

$$\le \sum_{l=0}^{\infty} C^2 \lambda^2 i^{d-1} j^d l^{2d-1} P(\rho = j) P(\rho = l)$$

$$= \sum_{l=0}^{\infty} C^2 \lambda^2 i^{d-1} j^d P(\rho = j) E\rho^{2d-1},$$

and a simple induction argument yields

$$\mu_{i,j}^{(n)} \le (C\lambda E\rho^{2d-1})^n P(\rho = j) i^{d-1} j^d. \tag{3.19}$$

Note that at this stage of the calculations in the previous theorem, we had the term $E\rho^{2d}$ in (3.10) and (3.11) instead of the term $E\rho^{2d-1}$ as in (3.19).

Hence, from (3.19),

$$\mu_i = \sum_{n=0}^{\infty} \sum_{j=0}^{\infty} \mu_{i,j}^{(n)}$$

$$\le i^{d-1} \sum_{n=0}^{\infty} (C\lambda E\rho^{2d-1})^n \sum_{j=1}^{\infty} j^d P(\rho = j).$$

Thus, if $C\lambda E\rho^{2d-1} < 1$, i.e., if λ is sufficiently small, we have that the expected number of balls which make up the boundary of the component of an arbitrary ball of radius i is finite. In particular, the number of boundary balls is finite a.s. From Corollary 3.1 we see that if a component is unbounded, then also the number of boundary balls must be infinite. We conclude that all components are bounded a.s.   □

REMARK: It follows from Theorem 3.3 that if $E\rho^{2d-1} < \infty$ then for all positive λ small enough, there are no unbounded occupied components a.s. We complement this result by showing that under the weakest possible condition, namely $P(\rho = 0) < 1$, we do get unbounded components when λ is large enough. To see this, choose $\epsilon > 0$ so that $P(\rho > \epsilon) > \epsilon$ and $1 - \epsilon > p_c(d)$ is the critical probability for independent site percolation in d dimensions. Next we choose $\delta > 0$ so small that if we partition the space by cubes with side length δ, any two points in neighbouring cubes are at a distance at most 2ϵ from each other. Next we choose N so large that$(1 - (1 - \epsilon)^N)(1 - \epsilon) > p_c(d)$, and finally we choose λ so large that the probability to have at least N Poisson points in a cube with side length δ is at least $1 - \epsilon$.

We call a cube open if it contains at least one point of the point process with a ball of radius at least ϵ. The choice of our parameters implies that the probability that a cube is open is larger than $p_c(d)$. Also, distinct cubes are independently open or closed. Identifying the cubes with the vertices of the d-dimensional integer lattice, we see that the union of all open cubes contains an unbounded component. Thus the Boolean model percolates.

The above two theorems have an interesting consequence:

Corollary 3.2 *If the radius random variable ρ of a Poisson Boolean model (X, ρ, λ) on $\mathbb{R}^d$, for $d \geq 2$, satisfies*

(i) $E\rho^{2d-1} < \infty$,

(ii) $E\rho^{2d} = \infty$

then there exists λ_0 such that, for all $0 < \lambda < \lambda_0$, with probability 1 no component contains an infinite number of Poisson points, whereas the expected number of Poisson points in the component containing the origin is infinite.

REMARK: Such a dichotomy does not occur in the standard percolation models on the discrete lattice.

3.4 Critical densities

In the previous section we noticed that depending on the density of the underlying Poisson process, the Boolean model is either subcritical – that is, the occupied component of the origin contains a finite number of Poisson points almost surely – or supercritical – that is, the occupied component of the origin contains an infinite number of Poisson points with positive probability. We also noticed the existence of two other phases of the Boolean model; the expected number of Poisson points in the occupied component containing the origin is finite in one such phase and infinite in the other phase. To formalise this *phase transition* we define the *critical densities*

$$\lambda_\# := \inf\{\lambda : P_\lambda\{X(W) = \infty\} > 0\} \tag{3.20}$$

and

$$\lambda_N := \inf\{\lambda : E_\lambda X(W) = \infty\}. \tag{3.21}$$

For a Poisson Boolean model (X, ρ, λ), Theorem 3.2 states that $\lambda_N > 0$ if and only if $E\rho^{2d} < \infty$, while if $E\rho^{2d-1} < \infty$ then Theorem 3.3 states that $\lambda_\# > 0$. In Corollary 3.2 we showed that if $E\rho^{2d-1} < \infty$ and $E\rho^{2d} = \infty$ then we have $0 = \lambda_N < \lambda_\#$.

In Examples 2.1 and 2.4 , we have used another notion of the size of the component, namely $d(W) = \sup\{d(x, y) : x, y \in W\}$. According to this notion of the size of W, we have the critical densities

$$\lambda_c := \inf\{\lambda : P_\lambda\{d(W) = \infty\} > 0\} \tag{3.22}$$

and

$$\lambda_D := \inf\{\lambda : E_\lambda d(W) = \infty\}. \tag{3.23}$$

Another notion of size which is very natural is the Lebesgue measure $\ell(W)$ of W. This leads to the critical densities

$$\lambda_H := \inf\{\lambda : P_\lambda\{\ell(W) = \infty\} > 0\} \tag{3.24}$$

and

$$\lambda_T := \inf\{\lambda : E_\lambda \ell(W) = \infty\}. \tag{3.25}$$

We remark here that all these critical densities depend on the underlying distribution of the radius random variable ρ. Thus if there is any scope for confusion we will write $\lambda_H(\rho)$ instead of λ_H to emphasise the underlying radius random variable. In a similar fashion, we express the dependence on ρ for the other critical densities whenever there is any chance of confusion. Our first concern is to show that the notion of size does not affect the critical densities when ρ is bounded. We show

Theorem 3.4 *In a Poisson Boolean model (X, ρ, λ) with*

$$0 < \rho \le R \ a.s. \ for \ some \ R > 0, \tag{3.26}$$

we have (a) $\lambda_\# = \lambda_c = \lambda_H$ and (b) $\lambda_N = \lambda_D = \lambda_T$.

Proof First suppose $\lambda > \lambda_\#$. Then, for some $\delta > 0$, $P_\lambda\{X(W) = \infty\} = \delta > 0$. Now for every $m > 0$ the box $B_m = [-m, m]^d$ contains at most a finite number of Poisson points a.s., thus $P_\lambda\{X(W \cap B_m^c) = \infty\} = \delta > 0$. But $X(W \cap B_m^c) = \infty$ implies that $X(W \cap B_m^c) > 0$; i.e., $d(W) \geq m$. Hence we have $P_\lambda\{d(W) \geq m\} = \delta > 0$. This being true for all $m > 0$, we have $\lambda \geq \lambda_c$. To show that $\lambda_\# \leq \lambda_c$, we note that $d(W) \leq 2RX(W)$, where R is as in (3.26). Thus $X(W) < \infty$ implies $d(W) < \infty$. This proves $\lambda_\# = \lambda_c$.

To show $\lambda_c = \lambda_H$, we observe that $\ell(W) \leq (d(W))^d$; i.e., if $d(W) < \infty$, then $\ell(W) < \infty$. This shows that $\lambda_c \leq \lambda_H$.

To show the reverse inequality we distinguish between two cases.

CASE 1: Suppose there exists some $\eta > 0$ such that $\rho \geq \eta$ a.s. If for some integer N, $d(W) \geq N$, then there must be at least $N/2R$ disjoint balls in the component W and so $\ell(W) \geq (N/2R)2\pi_d\eta^d$. Thus $\lambda_c \geq \lambda_H$.

CASE 2: Suppose there does not exist any η with $\rho \geq \eta$ a.s. Since $\rho > 0$ a.s. and $P(\rho = 0) = 0$, for any $\beta > 0$, we can find $0 < \alpha < \beta$ and r_0 such that

$$P(\rho < r_0) = \alpha < \beta. \tag{3.27}$$

Now take $\lambda > \lambda_c = \lambda_c(\rho)$ and choose $\beta = (\lambda - \lambda_c)/\lambda$. Let $\alpha < \beta$ satisfy (3.27) for this choice of β. Define $\bar{\lambda}$ by $\alpha = (\lambda - \bar{\lambda})/\lambda$. Since $0 < \alpha < \beta$, we have $\lambda_c < \bar{\lambda} < \lambda$. Thus, if we set $\mu = \lambda - \bar{\lambda}$, we have

$$P(\rho < r_0) = \frac{\mu}{\lambda}. \tag{3.28}$$

Since we shall use the technique of this proof more than once, we present the main idea of the proof before going into the details. We shall decompose the process (X, ρ, λ) into two independent processes such that (X, ρ, λ) is the superposition of these two processes. One of these processes will be chosen such that it has density $\bar{\lambda}$ with an associated radius distribution ρ_1 which is equivalent in law to ρ given $\rho \geq r_0$. The other process is now determined from the choice of the first process and, as we shall see shortly, it turns out to be a process with density μ and radius random variable ρ_2 which is equivalent in law to ρ given $\rho < r_0$. Now observe that the first process with density $\bar{\lambda}$ and radius ρ_1 'dominates' a process with density $\bar{\lambda}$ and radius ρ, thereby guaranteeing that the first process is supercritical in terms of the diameter of the occupied cluster. However, ρ_1 is bounded below by $r_0 > 0$ and thus by Case 1, this process is also supercritical in terms of the Lebesgue measure of the occupied cluster. Hence the superposition of this process with any other process will remain supercritical. In particular (X, ρ, λ), which is one such superposition, is supercritical in terms of the volume of the cluster. We now present the details of this idea.

Introduce two independent Poisson processes X_1 and X_2 on the same probability space as our Boolean model with densities $\bar{\lambda}$ and μ, respectively. Also, let ρ_1 and ρ_2 be two positive random variables with distributions given by

$$P(\rho_1 \geq r) = P(\rho \geq r | \rho \geq r_0)$$

$$= \begin{cases} \dfrac{P(\rho \geq r)}{P(\rho \geq r_0)} & \text{for } r \geq r_0 \\ 1 & \text{for } r < r_0, \end{cases} \tag{3.29}$$

and

$$P(\rho_2 \geq r) = P(\rho \geq r | \rho < r_0)$$

$$= \begin{cases} \dfrac{P(r \leq \rho < r_0)}{P(\rho < r_0)} & \text{for } r < r_0 \\ 0 & \text{for } r \geq r_0. \end{cases} \tag{3.30}$$

Now consider the Boolean models $(X_1, \rho_1, \bar{\lambda})$ and (X_2, ρ_2, μ). The superposition of these two Boolean models is a Boolean model with density $\bar{\lambda} + \mu = \lambda$. Moreover, for all x, $P(x \in X_1 | x \in X_1 * X_2) = (\lambda - \mu)/\lambda$, so the radius random variable associated with the superposed model of density λ is ρ. Let (X', ρ, λ) denote this superposed model. If W_1 and W' denote the occupied components of the origin in the Boolean models $(X_1, \rho_1, \bar{\lambda})$ and (X', ρ, λ), respectively, then

$$W_1 \subseteq W'. \tag{3.31}$$

Now let $(\bar{X}, \rho, \bar{\lambda})$ be a Poisson Boolean model independent of all the random quantities defined as yet, and let $\overline{W}$ denote the occupied component of the origin in this Boolean model. From (3.29) we have $P(\rho_1 \geq r) \geq P(\rho \geq r)$, which yields, by a coupling argument,

$$P_{\bar{\lambda}}(d(\overline{W}) = \infty) \leq P_{\bar{\lambda}}(d(W_1) = \infty). \tag{3.32}$$

But $\bar{\lambda} \geq \lambda_c(\rho)$; i.e.,

$$P_{\bar{\lambda}}(d(\overline{W}) = \infty) > 0, \tag{3.33}$$

so from (3.32) we have $P_{\bar{\lambda}}(d(W_1) = \infty) > 0$. However, from (3.29), $\rho_1 \geq r_0 > 0$ a.s., so the first case of this proof applied to the Boolean model $(X_1, \rho_1, \bar{\lambda})$ yields from (3.33)

$$P_{\bar{\lambda}}(\ell(W_1) = \infty) > 0.$$

Thus from (3.31), we have $P_\lambda(\ell(W') = \infty) > 0$, i.e., $\lambda \geq \lambda_H$. This proves (a) of the theorem.

To prove (b), it is easy to observe that the proofs in both the first and second case above go through when we consider λ_T instead of λ_H and λ_D instead of λ_c and take expectations instead of probabilities. This yields $\lambda_D = \lambda_T$.

Thus to complete the proof of the theorem we need to show that $\lambda_N = \lambda_T$. For this we observe that $\ell(W) \leq \pi_d R^d X(W)$, where π_d is, as usual, the d-dimensional volume of a unit ball. Thus if $E_\lambda X(W) < \infty$ then $E_\lambda(\ell(W)) < \infty$. This shows that $\lambda_N \leq \lambda_T$. To show $\lambda_N \geq \lambda_T$, we again consider two cases.

CASE 1: Suppose there exists $\eta > 0$ such that $\rho \geq \eta$ a.s. We partition $\mathbb{R}^d$ by the integer lattice $\mathbb{Z}^d$, and let I be a cell in this lattice. Let X_{I^c} denote the realisation of the Poisson process X outside the cell I. Let W_{I^c} denote the occupied component containing the origin in the 'Boolean model' (X_{I^c}, ρ, λ). (Note here that if I contains the origin, then for W_{I^c} to be non-empty, there must be a Poisson point outside I whose associated ball covers the origin.) Let $\delta(x)$ denote the (random) Euclidean distance from the point x to W_{I^c}. For all cells I at a distance at least R from the origin, we have

$$E_\lambda(X(W \cap I)|(X_{I^c}, \rho, \lambda))$$

$$= \sum_{k=1}^\infty k P_\lambda\{X(W \cap I) = k|(X_{I^c}, \rho, \lambda)\}$$

$$\leq \sum_{k=1}^\infty k P_\lambda\{X(I) \geq k \text{ and at least one of these } k \text{ points}$$

$$\text{has a ball which intersects } W_{I^c}|(X_{I^c}, \rho, \lambda)\}. \qquad (3.34)$$

Of course, these inequalities and all the subsequent inequalities in this proof which use conditional probability or expectation are 'almost sure' statements. To calculate the sum in the last inequality of (3.34) we observe that for $k \geq 2\lambda - 1$,

$$P_\lambda(X(I) \geq k) = e^{-\lambda} \sum_{n=k}^\infty \frac{\lambda^n}{n!}$$

$$\leq e^{-\lambda} \frac{\lambda^k}{k!} \left\{ \frac{1}{1 - \lambda/(k+1)} \right\}$$

$$\leq 2P(X(I) = k). \qquad (3.35)$$

Using the independence of the Poisson process X and the radius distribution, we have from (3.34) and (3.35)

$$E_\lambda(X(W \cap I)|(X_{I^c}, \rho, \lambda))$$

$$\leq \sum_{k=1}^{\infty} k P(X(I) \geq k) \left\{ 1 - \left(\int_I P(\rho < \delta(x_1)) dx_1 \right) \times \cdots \right.$$

$$\left. \times \left(\int_I P(\rho < \delta(x_k)) dx_k \right) \right\}$$

$$\leq \sum_{k=1}^{\lfloor 2\lambda-1 \rfloor} k P(X(I) \geq k) \left\{ k - k \left(\int_I P(\rho < \delta(x)) dx \right) \right\}$$

$$+ \sum_{k=\lfloor 2\lambda-1 \rfloor+1}^{\infty} 2k P(X(I) = k) \left\{ k - k \left(\int_I P(\rho < \delta(x)) dx \right) \right\}$$

$$\leq (2\lambda - 1)^2 \sum_{k=1}^{\lfloor 2\lambda-1 \rfloor} \left(\int_I P(\rho \geq \delta(x)) dx \right)$$

$$+ 2 \sum_{k=\lfloor 2\lambda-1 \rfloor+1}^{\infty} \frac{1}{k!} k^2 e^{-\lambda} \lambda^k \left(\int_I P(\rho \geq \delta(x)) dx \right)$$

$$= C(\lambda) \int_I P(\rho \geq \delta(x)) dx, \tag{3.36}$$

for some positive constant $C(\lambda) > 0$. Here we have used Proposition 1.2 at the first inequality and the inequality $1 - \prod_{i=1}^{n} a_i \leq \sum_{i=1}^{n} (1 - a_i)$ for $0 \leq a_1, a_2, \ldots, a_n \leq 1$ at the second inequality. Since every ball has a radius of at least η, a ball centred in a cell will cover a d-dimensional volume of at least $\min\{1, (1/2^d)\pi_d\eta^d\}$ inside the cell. Thus we have

$$E_\lambda(\ell(W \cap I)|(X_{I^c}, \rho, \lambda)) \geq \int_I v \exp(-\lambda)\lambda(P(\rho \geq \delta(x)) dx, \tag{3.37}$$

where $v = \min\{1, (1/2^d)\pi_d\eta^d\}$. So, from (3.36) and (3.37), for any cell I at a distance at least R from the origin, we obtain

$$E_\lambda(X(W \cap I)|(X_{I^c}, \rho, \lambda))$$

$$\leq \frac{C(\lambda)}{v \exp(-\lambda)\lambda} E_\lambda(\ell(W \cap I)|(X_{I^c}, \rho, \lambda)). \tag{3.38}$$

For a cell I at a distance less than R from the origin we have the trivial bound $E_\lambda X(W \cap I) \leq E_\lambda X(I) = \lambda$. Now taking expectations on both sides of (3.38) and summing over all cells I of the lattice we have

$$E_\lambda X(W) \leq \lambda(2R)^d + \frac{C(\lambda)}{\nu\lambda \exp(-\lambda)} E\ell(W).$$

This shows that $\lambda_T \leq \lambda_N$ in this case.

CASE 2: Suppose there does not exist any $\eta > 0$ such that $\rho \geq \eta$ a.s. As in case (ii) of part (a), for $\lambda < \lambda_T$ and $\beta = (\lambda_T - \lambda)/\lambda_T$, there exist $0 < \alpha < \beta$ and r_0 with $P(\rho < r_0) = \alpha$. Let $\bar{\lambda} > \lambda$ be such that $\alpha = (\bar{\lambda} - \lambda)/\bar{\lambda}$. Since $0 < \alpha < \beta$, we have $\lambda < \bar{\lambda} < \lambda_T$. Thus, if we set $\mu = \bar{\lambda} - \lambda$,

$$P(\rho < r_0) = \frac{\mu}{\lambda + \mu}. \tag{3.39}$$

Now let (X_1, ρ_1, λ) and (X_2, ρ_2, μ) be two independent Poisson Boolean models defined on the same probability space, where ρ_1 and ρ_2 are chosen as in (3.29) and (3.30). As in part (a) a little calculation shows that the superposition of these Boolean models is another Boolean model with density $\bar{\lambda}$ and radius random variable ρ. Let $(X', \rho, \bar{\lambda})$ be this superposed model. If W_1 and W' denote, as before, the occupied components of the origin in (X_1, ρ_1, λ) and $(X', \rho, \bar{\lambda})$ respectively, then we have

$$W_1 \subseteq W'. \tag{3.40}$$

Now let $(\bar{X}, \rho, \bar{\lambda})$ be a Poisson Boolean model independent of all the random quantities defined as yet, and let $\overline{W}$ denote the occupied component containing the origin in this model. Since $\bar{\lambda} < \lambda_T(\rho)$, we have $E_\lambda(\ell(\overline{W})) < \infty$. But $(\bar{X}, \rho, \bar{\lambda})$ and $(X', \rho, \bar{\lambda})$ are equivalent in law, so $E_{\bar{\lambda}}(\ell(W')) < \infty$. Thus, by (3.40), $E_{\bar{\lambda}}(\ell(W_1)) \leq E_{\bar{\lambda}}(\ell(W')) < \infty$. But $\rho_1 \geq r_0$ a.s., so by Case 1 we have $E_{\bar{\lambda}} X(W) < \infty$. Since $\lambda < \bar{\lambda}$, we have by a coupling argument $E_\lambda X(W) < \infty$ and consequently $\lambda \leq \lambda_N$. This completes the proof of the theorem.   □

Before we end this section we introduce another critical density based on the crossing probabilities introduced in Example 2.2. Recall the definition of $\sigma((n, 3n, \ldots, 3n), \lambda, 1)$ as the probability of the existence of an occupied crossing in the shortest direction of the rectangle $[0, n] \times [0, 3n] \times \cdots \times [0, 3n]$. Since the size of the rectangle increases in n in all directions, we do not have monotonicity of $\sigma((n, 3n, \ldots, 3n), \lambda, 1)$ in n. However, we can define the following:

$$\lambda_S = \lambda_S(\rho) = \inf\{\lambda : \limsup_{n\to\infty} \sigma((n, 3n, \ldots, 3n), \lambda, 1) > 0\}. \tag{3.41}$$

Proposition 3.2   *In any Poisson Boolean model we have* $\lambda_S \leq \lambda_c$.

Proof For ease of notation we restrict ourselves to two dimensions. The proof for higher dimensions proceeds along the same lines.

Consider the box $B_n = [-n, n] \times [-n, n]$. If $\lambda > \lambda_c$, then $P_\lambda\{d(W) = \infty\} > 0$ and so at least one of the following events occurs with probability at least $\frac{1}{4} P_\lambda\{d(W) = \infty\}$:

   (i) there is an L–R occupied crossing of the rectangle $[n, 3n] \times [-3n, 3n]$,
  (ii) there is an L–R occupied crossing of the rectangle $[-3n, -n] \times [-3n, 3n]$,
 (iii) there is a T–B occupied crossing of the rectangle $[-3n, 3n] \times [n, 3n]$,
  (iv) there is a T–B occupied crossing of the rectangle $[-3n, 3n] \times [-3n, -n]$.

This implies by translation and rotation invariance of the model that $\sigma((2n, 6n), \lambda, 1) > \frac{1}{4} P_\lambda(d(W) = \infty)$. This being true for all n, the proposition follows. □

3.5 Equality of the critical densities

In this section we show that in dimensions 2 or more, if the radius random variable is bounded, then the critical densities λ_c, λ_D and λ_S are all equal. Note that some condition on the radius random variable is necessary, because it follows from Corollary 3.2 that the result cannot be true in general. The proof we present uses a lattice approximation and the scaling relation Proposition 2.11 to show first that in case the balls are all of a fixed size, the equality holds. In case the radius random variable ρ takes on finitely many distinct values, the approximating lattice we need is a multi-parametric one. Finally, a general ρ is approximated from below and above by random variables U_n and V_n, each of which take finitely many values and this approximation yields the desired equality.

Theorem 3.5 *For a Poisson Boolean model (X, ρ, λ) on $\mathbb{R}^d$, $d \geq 2$, with ρ bounded almost surely, we have $\lambda_c(\rho) = \lambda_D(\rho) = \lambda_S(\rho)$.*

Proof

CASE 1: First we consider the case when ρ is a fixed constant. It can be easily seen that Proposition 2.11 holds for both λ_D and λ_S in addition to that for λ_c as stated. As such, it suffices to consider $\rho \equiv 1$ and prove the equality of the critical densities in this case.

Consider a discrete percolation model described as follows. Let $\mathcal{V}_n$ be the set of vertices of the lattice $\mathbb{L}_n := ((1/n)\mathbb{Z})^d$ and for a vertex $v = (v_1, v_2, \ldots, v_d) \in \mathcal{V}_n$, let $K_n(v) = [v_1 - (1/2n), v_1 + (1/2n)) \times \cdots \times [v_d - (1/2n), v_d + (1/2n))$ be the cell containing v. Let $\mathcal{G}_n$ be the graph with vertices $\mathcal{V}_n$ and edges constructed by joining all pairs of vertices v and w of V_n with $d(K_n(v),$ $K_n(w)) \leq 2$, where $d(A, B) = \inf\{d(a, b) : a \in A, b \in B\}$ for any two regions $A, B \subseteq \mathbb{R}^d$ and $d(a, b)$ being the Euclidean distance between a and b. A vertex of this graph is *open* with probability p and *closed* with probability $1 - p$ independently of all other vertices. This graph is clearly isomorphic to the one-parametric site-percolation graph described in the paragraphs preceding Theorem 1.7 for a suitable L. Returning to our graph $\mathcal{G}_n$, we may define the critical values $p_c(\mathcal{G}_n)$, $p_T(\mathcal{G}_n)$ and $p_S(\mathcal{G}_n)$ as in Section 1.2 and Theorem 1.7

$$p_c(\mathcal{G}_n) = p_T(\mathcal{G}_n) = p_S(\mathcal{G}_n). \tag{3.42}$$

We now incorporate the continuum model $(X, 1, \lambda)$ in this site-percolation model on $\mathcal{G}_n$. On $\mathbb{R}^d$ we place the graph $\mathcal{G}_n$ and for any $x \in \mathbb{R}^d$, let $v(x)$ denote the vertex v in $\mathcal{G}_n$ such that $x \in K_n(v)$. A vertex v is open if $X(K_n(v)) \geq 1$ and it is closed if $X(K_n(v)) = 0$; i.e., a vertex v is open if and only if there is at least one Poisson point in the cell containing v. Clearly, $P_\lambda(v$ is open$) = 1 - \exp(-\lambda/n^d) = p_n(\lambda)$ (say) and $P_\lambda(v$ is closed$) = 1 - p_n(\lambda)$ for every $v \in V_n$; moreover, v is open or closed independently of other vertices. This is indeed the same site-percolation problem as described in the previous paragraph and is governed by the same set of critical values.

If x and y are two Poisson points of X such that $d(x, y) \leq 2$, i.e. $S(x, 1) \cap S(y, 1) \neq \emptyset$, then either $v(x) = v(y)$ or $v(x)$ and $v(y)$ are adjacent in the sense that there is an edge in $\mathcal{G}_n$ connecting $v(x)$ and $v(y)$. Thus if there is an unbounded occupied cluster in the continuum model, then there is an unbounded open cluster in its approximating site-percolation model on $\mathcal{G}_n$; i.e., if $\lambda > \lambda_c(1)$, then $p_n(\lambda) \geq p_H(\mathcal{G}_n)$. This shows that $p_n(\lambda_c(1)) \geq p_H(\mathcal{G}_n)$ for every $n \geq 1$. Since $p_n(\lambda)$ is increasing in λ, we can take the inverse of the function p_n and restate the inequality we just obtained as

$$\lambda_c(1) \geq p_n^{-1}(p_H(\mathcal{G}_n)). \tag{3.43}$$

Now we scale the radius of the continuum model by a factor $l_n := 1 + \sqrt{d/n}$ and consider the model (X, l_n, λ). Notice that if v and w are two adjacent vertices in $\mathcal{G}_n$ then $\sup\{d(x, y) : x \in K_n(v), y \in K_n(w)\} \leq 2 + 2\sqrt{d/n} = 2l_n$, and so if v and w are two adjacent open vertices and x and y are two Poisson points in $K_n(v)$ and $K_n(w)$, respectively, then $d(x, y) \leq 2l_n$. Thus if there is an unbounded open cluster in $\mathcal{G}_n$, then there is an unbounded occupied cluster in (X, l_n, λ); i.e., if $p > p_c(\mathcal{G}_n)$ then $p_n^{-1}(p) \geq \lambda_c(l_n)$. This gives us $\lambda_c(l_n) \leq p_n^{-1}(p_c(\mathcal{G}_n))$, and in conjunction with (3.43) we have

$$\lambda_c(l_n) \leq p_n^{-1}(p_H(\mathcal{G}_n)) \leq \lambda_c(1). \tag{3.44}$$

A similar argument yields

$$\lambda_D(l_n) \le p_n^{-1}(p_T(\mathcal{G}_n)) \le \lambda_D(1) \tag{3.45}$$

and

$$\lambda_S(l_n) \le p_n^{-1}(p_S(\mathcal{G}_n)) \le \lambda_S(1). \tag{3.46}$$

Now the scaling relations (Proposition 2.11) and its equivalent version for the other critical densities imply that $\lambda_c(l_n) \to \lambda_c(1)$, $\lambda_T(l_n) \to \lambda_T(1)$ and $\lambda_S(l_n) \to \lambda_S(1)$ as $n \to \infty$. The equality of the critical densities now follows from this observation, the equality (3.42) and the inequalities (3.44), (3.45) and (3.46).

CASE 2: Now suppose ρ takes only k distinct values $r_1, \ldots, r_k$. Instead of the single-parameter site-percolation model we considered in the previous case, we now consider a graph $\mathcal{G}_n$ which consists of k layers as described in the paragraph preceding Theorem 1.7 for $k = 2$. The i-th layer consists of the vertices of the lattice $\mathbb{L}_n$ and the edges constructed by connecting any pair of vertices v and w in this layer if and only if $d(K_n(v), K_n(w)) \le 2r_i$. A vertex v in the i-th layer and a vertex w in the j-th layer $(i \ne j)$ are connected by an edge if $d(K_n(v), K_n(w)) \le r_i + r_j$. The graph $\mathcal{G}_n$ consists of all the vertices of the different layers and all the edges we have described. We define a site-percolation model where a vertex v of the i-th layer is open with probability p_i, where $p_1 + \cdots + p_k = 1$ independently of all other vertices in all the layers. As in Theorem 1.7, we get the equality (3.42) of the critical regions described by $p_c(\mathcal{G}_n)$, $p_T(\mathcal{G}_n)$ and $p_S(\mathcal{G}_n)$.

To connect the continuum model (X, ρ, λ) with this site-percolation model on the graph $\mathcal{G}_n$, we call a vertex v of the i-th layer open if and only if on an embedding of the i-th layer in $\mathbb{R}^d$, the cell $K_n(v)$ contains at least one Poisson point of X with an associated ball of radius r_i. A similar application of the scaling relations and comparison with this approximating graph $\mathcal{G}_n$ will yield the desired equality of the critical densities in this case.

CASE 3: Next we consider the case where the support of ρ is contained in an interval $[a, a + R]$ with $a, R > 0$. Let $n > 0$ and consider the set of points $\{a + k2^{-n}, k = 0, \ldots, \lfloor 2^n R \rfloor\}$ and define

$$k(n) = \max_{0 \le k \le \lfloor 2^n R \rfloor} \left\{ \frac{a + (k+1)2^{-n}}{a + k2^{-n}} \right\}.$$

It is easy to see that $\lim_{n \to \infty} k(n) = 1$. We define, on the same probability space as the Boolean model, the random variable V_n as follows: if

$a + k2^{-n} \leq \rho < a + (k+1)2^{-n}$, then we put $V_n = a + k2^{-n}$. Note that it follows from these definitions that

$$V_n \leq \rho \leq k(n)V_n. \tag{3.47}$$

It follows from (3.47) and Proposition 2.11 that

$$\lambda_c(k(n)V_n) = \lambda_c(V_n)k(n)^{-d}$$

and a simple coupling argument gives $\lambda_c(V_n) \geq \lambda_c(\rho)$. Putting these things together we find

$$\lambda_c(V_n) \geq \lambda_c(\rho) \geq k(n)^{-d}\lambda_c(V_n). \tag{3.48}$$

In a similar way we obtain

$$\lambda_D(V_n) \geq \lambda_D(\rho) \geq k(n)^{-d}\lambda_D(V_n) \tag{3.49}$$

and

$$\lambda_S(V_n) \geq \lambda_S(\rho) \geq k(n)^{-d}\lambda_S(V_n). \tag{3.50}$$

It follows from Case 2 that all critical densities for V_n are equal. Now note that $\lambda_c(V_n)$ is non-increasing in n, whence $\lim_{n\to\infty}\lambda_c(V_n)$ exists. Hence we have from (3.48), (3.49), (3.50) and the fact that $\lim_{n\to\infty}k(n) = 1$ that all of $\lambda_c(\rho)$, $\lambda_D(\rho)$ and $\lambda_S(\rho)$ are equal to $\lim_{n\to\infty}\lambda_c(V_n)$. This completes the proof for this case.

CASE 4:   Finally we remove all restrictions on ρ (apart from it being bounded). Let $\epsilon > 0$ and choose $a = a(\epsilon)$ so small that $P(\rho \leq a) \leq \epsilon$. Let ρ^a be a random variable with distribution equal to the conditional distribution of ρ conditioned on $\rho \geq a$. Similarly, let ρ_a be a random variable with distribution equal to the conditional distribution of ρ conditioned on $\rho < a$. Then we have by a simple coupling argument that $\lambda_c(\rho^a) \leq \lambda_c(\rho)$.

Consider two independent models (X_1, ρ^a, λ) and $(X_2, \rho_a, \lambda l)$, where l is chosen such that $l(1 + l)^{-1} = P(\rho \leq a)$; i.e.

$$l = \frac{P(\rho \leq a)}{P(\rho > a)} \leq \frac{\epsilon}{1 - \epsilon}. \tag{3.51}$$

The superposition of the two models is equivalent in law to a process $(X, \rho, \lambda(1 + l))$. Thus if $\lambda > \lambda_c(\rho^a)$, then certainly this superposition is supercritical and hence $\lambda(1 + l) > \lambda_c(\rho)$; i.e. $\lambda_c(\rho^a)(1 + l) \geq \lambda_c(\rho)$. Hence

$$|\lambda_c(\rho) - \lambda_c(\rho^a)| \leq l\lambda_c(\rho^a) \leq \frac{\epsilon}{1 - \epsilon}\lambda_c(\rho), \tag{3.52}$$

where we have used (3.51). Because $\lambda_c(\rho) < \infty$, we see that $\lambda_c(\rho^{a(\epsilon)}) \to \lambda_c(\rho)$ when ϵ tends to zero. In a similar fashion, we find that $\lambda_S(\rho^{a(\epsilon)}) \to \lambda_S(\rho)$ and $\lambda_D(\rho^{a(\epsilon)}) \to \lambda_D(\rho)$ when $\epsilon \to 0$. From Case 3 it follows that $\lambda_c(\rho^{a(\epsilon)}) = \lambda_D(\rho^{a(\epsilon)}) = \lambda_S(\rho^{a(\epsilon)})$ for all $\epsilon > 0$ and the proof is complete.

$\square$

3.6 Uniqueness

We already concluded in Chapter 2 that in the supercritical regime, i.e. if $\lambda > \lambda_c(\rho)$, then unbounded occupied components exist almost surely and that the number of such components is an almost sure constant (which could be infinity). As we shall see now, in the case of a Poisson Boolean model there can be at most one unbounded occupied component a.s. In Chapter 7, we will prove a uniqueness result for Boolean models driven by arbitrary point processes.

Theorem 3.6 *In a Poisson Boolean model* (X, ρ, λ), *there can be at most one unbounded occupied component a.s.*

The rest of this section is devoted to a proof of this result. It is quite involved and requires some preliminary results. We start with a combinatorial result which turns out to be of great value and will be used a couple of times throughout the book.

Lemma 3.2 *Let S be a set and let R be a non-empty finite subset of S. Suppose that*

(a) *for all $r \in R$, we have a family $(C_r^{(1)}, C_r^{(2)}, C_r^{(3)})$ of disjoint non-empty subsets (which we shall call* branches) *of S, not containing r, and* $\operatorname{card}(C_r^{(i)}) \geq K$, *for all i and r, where* $\operatorname{card}(\cdot)$ *denotes the cardinality of a set,*

(b) *for all $r, r' \in R$, one of the following events occurs, writing C_r for $\cup_{i=1}^3 C_r^{(i)}$,*

  (i) $(\{r\} \cup C_r) \cap (\{r'\} \cup C_{r'}) = \emptyset$

  (ii) *there exist i, j such that $C_r^{(i)} \supset \{r'\} \cup C_{r'} \backslash C_{r'}^{(j)}$ and $C_{r'}^{(j)} \supset \{r\} \cup C_r \backslash C_r^{(i)}$.*

Then $\operatorname{card}(S) \geq K(\operatorname{card}(R) + 2) + \operatorname{card}(R)$, *where* $\operatorname{card}(\cdot)$ *denotes the cardinality of a set.*

Proof First we claim that there exist $r_0 \in R$ and $i_0 \in \{1, 2, 3\}$ such that

$$C_{r_0}^{(i_0)} \cap R = \emptyset. \tag{3.53}$$

To see this choose any $r_1 \in R$ and $i_1 \in \{1, 2, 3\}$. If $C_{r_1}^{(i_1)} \cap R = \emptyset$ we are done. If not, then there is an element $r_2 \in C_{r_1}^{(i_1)} \cap R$. For this r_2 we have that

$$r_2 \in (\{r_1\} \cup C_{r_1}) \cap (\{r_2\} \cup C_{r_2}).$$

Hence it must be the case that for some j_1 and j_2 we have

$$\{r_1\} \cup C_{r_1} \backslash C_{r_1}^{(j_1)} \subset C_{r_2}^{(j_2)}$$

and

$$\{r_2\} \cup C_{r_2} \backslash C_{r_2}^{(j_2)} \subset C_{r_1}^{(j_1)}. \tag{3.54}$$

Using the fact that $r_2 \in C_{r_1}^{(i_1)}$ we conclude from (3.54) that $j_1 = i_1$. Hence $C_{r_2} \backslash C_{r_2}^{(j_2)} \subset C_{r_1}^{(i_1)}$. This means that there must be some $k_2 \neq j_2$ for which $C_{r_2}^{(k_2)} \subset C_{r_1}^{(i_1)}$. It follows that $C_{r_2}^{(k_2)} \cap R \subset C_{r_1}^{(i_1)} \cap R$ and this inclusion is strict because r_2 is an element of the right-hand side but not of the left-hand side. Hence

$$\operatorname{card}(C_{r_2}^{(k_2)} \cap R) < \operatorname{card}(C_{r_1}^{(i_1)} \cap R).$$

We can repeat this procedure only finitely many times because R is a finite set. It follows that eventually we find r_0 and i_0 as in (3.53).

Next we remove r_0 and $C_{r_0}^{(i_0)}$ from our set. Thus we put $R' := R \backslash \{r_0\}$ and $S' := S \backslash C_{r_0}^{(i_0)}$. We claim that for S' and R' properties (a) and (b) still hold, where R is replaced by R' and S by S'. To prove this claim, let $r \in R'$. If $C_{r_0} \cap C_r = \emptyset$, nothing has been changed in C_r. If not, then there are j and j_0 such that

$$\{r_0\} \cup C_{r_0} \backslash C_{r_0}^{(j_0)} \subset C_r^{(j)}$$

and

$$\{r\} \cup C_r \backslash C_r^{(j)} \subset C_{r_0}^{(j_0)}.$$

The element $\{r\}$ is still in S' and hence it follows from the last inclusion that $j_0 \neq i_0$. This means that at least two branches of $\{r\}$ are unaffected by the removal of $\{r_0\}$ and $C_{r_0}^{(i_0)}$. So for each $r \in R'$, only one branch, $C_r^{(1)}$ say, may have been changed into $C_r^{(1)} \backslash C_{r_0}^{(i_0)}$. If this happens, it implies that $C_{r_0}^{(i_0)} \subset C_r^{(1)}$ and hence there exist k and k_0 such that

$$\{r_0\} \cup C_{r_0} \backslash C_{r_0}^{(k_0)} \subset C_r^{(k)} \tag{3.55}$$

and

$$\{r\} \cup C_r \backslash C_r^{(k)} \subset C_{r_0}^{(k_0)}. \tag{3.56}$$

If $k \neq 1$ in (3.55), then k_0 has to be i_0. But then it would follow from (3.56) that $\{r\}$ has two branches in $C_{r_0}^{(i_0)}$, which is impossible by the observation above.

Thus $k = 1$ and $k_0 \neq i_0$. From (3.55) it then follows that $C_r^{(1)}$ contains at least one branch of $\{r_0\}$ different from $C_{r_0}^{(i_0)}$ and hence even after removing $\{r_0\}$ and $C_{r_0}^{(i_0)}$ it is still the case that $\mathrm{card}(C_r^{(1)}) \geq K$. This shows that (a) still holds. The inclusions in (b) remain true because for each $r \in R'$ we remove points in at most one branch of r'.

Finally, we repeat this procedure of taking away points in R and associated branches not containing points in R until only one element in R remains. This means that we do this step $\mathrm{card}(R) - 1$ times, and each time we take away at least $K + 1$ points. In the end, we are left with at least $3K + 1$ points, namely the remaining point in R together with its branches, each of which still contains at least K points. Hence the original set S contained at least $3K + 1 + (K + 1)(\mathrm{card}(R) - 1)$ points, proving the lemma. □

Here is the first step towards a uniqueness result:

Proposition 3.3 *In a Poisson Boolean model, the number of unbounded occupied components is equal to either zero, one or infinity almost surely.*

Proof It follows from the discussion on ergodicity in Chapter 2 that a Poisson Boolean model is ergodic (Proposition 2.6) and that the number of unbounded occupied components in such a model is an a.s. constant (Theorem 2.1).

First, we consider the case where the support of ρ is unbounded. The proof proceeds by contradiction, so we suppose that the number of unbounded occupied components is a.s. equal to $K \geq 2$, say. If B_n is the box $[-n, n]^d$ as usual, it is clear that for n large enough, there is a positive probability that all K unbounded occupied components have non-empty intersection with B_n and, in addition, $X(B_n) \geq 1$. Also, given any enumeration $\{x_1, x_2, \ldots\}$ of the points of X according to a fixed rule (for instance the absolute value), there is an index m such that the event $E := \{$all unbounded components intersect $B_n, x_m \in B_n\}$ has positive probability. (Note that neither n nor m are random.) From the fact that ρ has unbounded support, it follows that the event $E^* = E \cap \{\rho_m > 2n\sqrt{d}\}$ also has positive probability, where ρ_m is the radius of the ball centred at x_m. On E^*, however, the number of unbounded occupied components is equal to 1, because the ball $S(x_m, \rho_m)$ connects all K formerly unbounded components. This is the desired contradiction. Note that we have not used the fact that X is a Poisson process so far.

It remains to prove the lemma in the case where the support of ρ is bounded. The idea here is the same as in the previous case, but the procedure to connect together different components is a little more involved. First, we find $M > 0$ such that

(i) $P(\rho > M) = 0$,

(ii) $P(M - \eta < \rho \le M) > 0$, for any $\eta > 0$.

Now suppose (X, ρ, λ) admits $K \ge 2$ unbounded occupied components a.s. If we remove all the balls centred inside a box B, then the resulting configuration should contain at least K unbounded components a.s. Let, for $A \subset \mathbb{R}^d$, $C[A]$ denote the region $\cup_{x_i \in A} S(x_i, \rho_i)$; that is, $C[A]$ is the occupied region formed by points of X in A. Given a box B and $\epsilon > 0$, consider the event $A(B, \epsilon) := \{d(U, B) \le M - \epsilon$ for any unbounded occupied component U in $C[B^c]\}$. Partition the box into cubic cells with edge length $a > 0$ and let $C_a = \{G_1, \ldots, G_N\}$ denote the collection of all the cells which are adjacent to the boundary of B. Clearly, for a box B and $\epsilon > 0$, we can find $a = a(B, \epsilon) > 0$ and $\eta = \eta(a) > 0$ such that for any point $x \notin B$ with $d(x, B) \le M - (\epsilon/2)$, there exists a cell $G = G(x) \in C_a$ for which we have $\sup_{y \in G} d(x, y) \le M - 2\eta$. This means that, if we centre in each cell of C_a a ball with radius between $M - \eta$ and M, then the region $\{x \notin B : d(x, B) \le M - \epsilon/2\}$ will be completely covered by these balls.

Let $E = E(a, \eta)$ be the event that each cell in C_a contains at least one Poisson point with an associated ball of radius between $M - \eta$ and M. Since E depends on the configuration inside the box B and $A(B, \epsilon)$ depends on the configuration outside the box B, and the radii are independent of the Poisson process, we have

$$P(A(B, \epsilon) \cap E) = P(A(B, \epsilon))P(E).$$

If both $A(B, \epsilon)$ and E occur, then there is only one unbounded occupied component. Now $P(E) > 0$, so in order to arrive at a contradiction, we need to show that there exist a box B and an $\epsilon > 0$ such that $P(A(B, \epsilon)) > 0$. Since (X, ρ, λ) admits $K \ge 2$ unbounded occupied components, we can find a box B so large that, with positive probability, $d(U, B) < M$ for every unbounded component U of C. Also, the radius of any ball is at most M, so, with positive probability, $d(U, B) < M$ for every unbounded occupied component U in $C[B^c]$. Thus for this B we can find $\epsilon > 0$ such that $A(B, \epsilon)$ occurs with positive probability.

$\square$

Proof of Theorem 3.6   According to Proposition 3.3 it suffices to rule out the possibility of having infinitely many unbounded occupied components, so we again proceed by assuming the contrary and then derive a contradiction. Suppose there are infinitely many unbounded occupied components a.s. Define

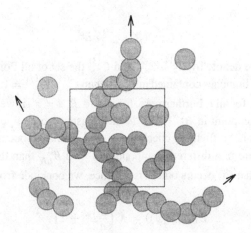

Figure 3.1. The branches of C' are the connected regions outside the box.

for each integer n and $z = (z_1, \ldots, z_d) \in \mathbf{Z}^d$

$$B_n^z := B_n^0 + (z_1, \ldots, z_d),$$

where of course $B_n^0 = [-n, n]^d$. (Note that in this notation, $B_n = B_n^0$.) As in the proof of Proposition 3.3 we can find an N such that $C[B_N^c]$ contains at least three unbounded occupied components which can be connected to each other via extra balls centred in B_N. It follows from this that for some N, the following event has positive probability η (say):

$E^0(N) = \{$there is an unbounded occupied component C' with the property that $C' \cap (B_N^0)^c$ contains at least three unbounded components and such that there is at least one Poisson point in $C' \cap B_N^0\}$.

We shall call the unbounded components in $C' \cap (B_N^0)^c$ *branches* (see Figure 3.1). Next we choose K very large, we shall see at the end of the proof exactly how large. Given K, we choose M so large that the following event has probability at least $\frac{1}{2}\eta$:

$E^0(N, M) := E^0(N) \cap \{$all three branches of B_N^0 contain at least K Poisson points in $B_{MN}^0 \setminus B_N^0\}$.

The events $E^z(N)$ and $E^z(N, M)$ are defined by translating $E^0(N)$ and $E^0(N, M)$ over the vector z. It follows that if R is the set

$$R := \{z \in \mathbf{Z}^d : B_{MN}^{2Nz} \subset B_{LN}^0, \ E^{2Nz}(N, M) \text{ occurs}\},$$

then, for any $L > 0$,

$$E(\text{card}(R)) \geq \tfrac{1}{4}\eta L^d. \tag{3.57}$$

For $z \in R$, if we denote by $C_z^{(1)}$, $C_z^{(2)}$ and $C_z^{(3)}$ the set of all Poisson points in each of its three branches contained in B_{MN}^{2Nz}, then $C_z^{(i)} \cap C_z^{(j)} = \emptyset$ for $i \neq j$, and $\text{card}(C_z^{(i)}) \geq K$ for all i. Furthermore, for $z, z' \in R$, $z \neq z'$, if we identify z and z' with a Poisson point in $B_N^{2Nz} \cap C'$ (which exists according to the definition of $E^0(N)$), it is not difficult to check that (i) in Lemma 3.2 occurs if the points of X in B_N^{2Nz} are in a different component of $C \cap B_{LN}^0$ than the points of X in $B_N^{2Nz'}$, and that (ii) occurs otherwise. Hence, we conclude from Lemma 3.2 and (3.57) that

$$E(X(B_{LN})) \geq K\left(\frac{\eta}{4}L^d + 2\right). \tag{3.58}$$

To see that this leads to a contradiction, note that

$$E(X(B_{LN})) = \lambda(2LN)^d. \tag{3.59}$$

Hence we find from (3.58) and (3.59) that for L large enough,

$$K\left(\frac{\eta}{4}L^d + 2\right) \leq \lambda(2LN)^d,$$

which gives the desired contradiction if we choose K large. $\qquad\square$

3.7 Exponential decay

We now prove a result which allows us to give bounds on the growth of the occupied cluster of the origin in the subcritical regime. Consider a Poisson Boolean model (X, ρ, λ) with $0 < \rho \leq R$. Recall that $\sigma((N_1, \ldots, N_d), \lambda, i) = P_\lambda$ {there exists an occupied crossing in the i-th direction of the rectangle $[0, N_1] \times \cdots \times [0, N_d]$}.

Lemma 3.3 *Consider a Boolean model with $\rho \leq R$ a.s. Let $\kappa_0 = (e3^d)^{-11^d-1}$. If $\sigma((3N_1, \ldots, 3N_{i-1}, N_i, 3N_{i+1}, \ldots, 3N_d), \lambda, i) < \kappa_0$ for all $i = 1, \ldots, d$ and for some $N_1, \ldots, N_d$ with $N_j > R$ for all $1 \leq j \leq d$, then, for all a sufficiently large, we have*

$$P_\lambda(d(W) > a) \leq C_1 \exp(-C_2 a) \tag{3.60}$$

and

$$P_\lambda(\ell(W) > a) \leq C_3 \exp(-C_4 a) \tag{3.61}$$

for some positive constants C_1, C_2, C_3 and C_4.

Proof Consider the integer lattice $\mathbf{Z}^d$ with vertices $(v_1, \ldots, v_d)$ and $(w_1, \ldots, w_d)$ adjacent if and only if $\max\{|v_i - w_i|, i = 1, \ldots, d\} = 1$. A vertex $z \in \mathbf{Z}^d$ is open if and only if there exists a connected component Λ of the covered region C of the Boolean model (X, ρ, λ) such that

$$\Lambda \cap ([z_1N_1, (z_1 + 1)N_1) \times \cdots \times [z_dN_d, (z_d + 1)N_d)) \neq \emptyset \qquad (3.62)$$

and

$$\Lambda \cap ([(z_1 - 1)N_1, (z_1 + 2)N_1) \times \cdots \\ \times [(z_d - 1)N_d, (z_d + 2)N_d))^c \neq \emptyset, \qquad (3.63)$$

where $N_1, N_2, \ldots, N_d$ are as in the lemma. The vertex z is closed if it is not open. Clearly we have constructed a dependent site percolation process where the state of a vertex z depends on the configuration of the Boolean model (X, ρ, λ) in the region

$$[(z_1 - 1)N_1 - R, (z_1 + 2)N_1 + R) \times \cdots \\ \times [(z_d - 1)N_d - R, (z_d + 2)N_d + R) \\ \subseteq [(z_1 - 2)N_1, (z_1 + 3)N_1) \times \cdots \\ \times [(z_d - 2)N_d, (z_d + 3)N_d).$$

Thus if z and z' are vertices of $\mathbf{Z}^d$ such that $\max\{|z_i - z_i'|; i = 1, \ldots, d\} \geq 5$ then the states of the vertices z and z' are independent of each other. Also observe that for a vertex z to be open the Boolean model must admit an occupied crossing in the i-th direction of a suitable translate of the rectangle $[0, 3N_1] \times \cdots \times [0, 3N_{i-1}] \times [0, N_i] \times [0, 3N_{i+1}] \times \cdots \times [0, 3N_d]$ for some $i = 1, \ldots, d$. Thus, by the hypothesis of the lemma,

$$p := P_\lambda(z \text{ is open})$$
$$\leq 2 \sum_{i=1}^{d} \sigma((3N_1, \ldots, 3N_{i-1}, N_i, 3N_{i+1}, \ldots, 3N_d), \lambda, i)$$
$$\leq 2d\kappa_0. \qquad (3.64)$$

Now suppose $d(W) \geq a$. Since, for any $z \in \mathbf{Z}^d$,

$$d(W \cap ([z_1N_1, (z_1 + 1)N_1) \times \cdots \times [z_dN_d, (z_d + 1)N_d))) \\ \leq d([z_1N_1, (z_1 + 1)N_1) \times \cdots \times [z_dN_d, (z_d + 1)N_d)) \\ \leq \sqrt{N_1^2 + \cdots + N_d^2}$$

and

$$d(W \cap ([(z_1 - 1)N_1, (z_1 + 2)N_1) \times \cdots \times [(z_d - 1)N_d, (z_d + 2)N_d)))$$
$$\leq d(([(z_1 - 1)N_1, (z_1 + 2)N_1) \times \cdots \times [(z_d - 1)N_d, (z_d + 2)N_d))$$
$$\leq 3\sqrt{N_1^2 + \cdots + N_d^2},$$

there must be at least $a(N_1^2 + \cdots + N_d^2)^{-1/2}$ vertices $z \in Z^d$ such that (3.62) holds. Also, if a is such that $a > 3\sqrt{N_1^2 + \cdots + N_d^2}$, then each of the $a(N_1^2 + \cdots + N_d^2)^{-1/2}$ vertices z in Z^d satisfies (3.63). Moreover, for $z = 0$, the origin, both (3.62) and (3.63) hold whenever $d(W) \geq a$, thus we have that $\text{card}(C') \geq a(N_1^2 + \cdots + N_d^2)^{-1/2}$ whenever $d(W) \geq a, a \geq 3\sqrt{N_1^2 + \cdots + N_d^2}$, where C' is the open cluster of the origin in the lattice Z^d (with the adjacency structure given above).

A similar argument yields that for $a \geq 3^d N_1 \cdots N_d$, whenever $\ell(W) \geq a$, we must have $\text{card}(C') \geq a(N_1 \cdots N_d)^{-1}$.

Thus for a large enough,

$$P_\lambda(d(W) \geq a) \leq P(\text{card}(C') \geq a(N_1^2 + \cdots + N_d^2)^{-1/2}) \tag{3.65}$$

and

$$P_\lambda(\ell(W) \geq a) \leq P(\text{card}(C') \geq a(N_1 \cdots N_d)^{-1}). \tag{3.66}$$

Now given a set S_n of n vertices of Z^d, since the states of two vertices z and z' are independent whenever $\max\{|z_i - z_i'|; i = 1, \ldots, d\} \geq 5$, we have that S_n must contain at least $n11^{-d}$ vertices whose states are independent of each other. Thus

$$P_\lambda(\text{all vertices of } S_n \text{ are open}) \leq p^{n/11^d} \tag{3.67}$$

where p is as defined in (3.64).

Hence from (3.67) we have

$$P_\lambda(\text{card}(C') = n) = \sum_{S_n} P_\lambda(C' = S_n)$$
$$\leq b_n p^{n/11^d}, \tag{3.68}$$

where the sum is over all connected sets S_n of n vertices of Z^d containing the origin and b_n is the total number of such sets S_n. Combining (3.68) with (3.65) and using the estimate $b_n \leq (e3^d)^n$ (Kesten 1982, Lemma 5.1), we have

$$
P_\lambda(d(W) \geq a) \leq \sum_{n \geq a(N_1^2 + \cdots + N_d^2)^{-1/2}} P_\lambda(\text{card}(C') = n)
$$

$$
\leq \sum_{n \geq a(N_1^2 + \cdots + N_d^2)^{-1/2}} b_n\, p^{n/11^d}
$$

$$
\leq \sum_{n \geq a(N_1^2 + \cdots + N_d^2)^{-1/2}} (e3^d)^n\, p^{n/11^d}
$$

$$
< C_1 \exp(-C_2 a), \tag{3.69}
$$

where

$$
C_1 = (1 - e3^d (e3^d \kappa_0)^{11^{-d}})^{-1}
$$

and

$$
C_2 = -\frac{1 + d \log 3}{\sqrt{N_1^2 + \cdots + N_d^2}} - \frac{1 + d \log 3}{11^d \sqrt{N_1^2 + \cdots + N_d^2}} - \frac{\log \kappa_0}{11^d \sqrt{N_1^2 + \cdots + N_d^2}}
$$

which is positive by our choice of κ_0. This proves (3.60). Combining (3.68) with (3.66) and performing similar calculations as in (3.69) yields (3.61). This proves the lemma. □

3.8 Continuity of the critical density and the percolation function

The Boolean model (X, ρ, λ) has two parameters λ and ρ. In this section we investigate to what extent quantities like $\lambda_c(\rho)$ and $\theta_\rho(\lambda)$ are continuous with respect to these parameters. For continuity with respect to ρ we need to choose a notion of convergence, and we shall use weak convergence throughout. The first question we want to answer is whether or not $\lambda_c(\rho_k)$ converges to $\lambda_c(\rho)$ when $\rho_k \Rightarrow \rho$, where '$\Rightarrow$' denotes weak convergence. In fact we already know that this can not be true in general. In Proposition 3.1 we already proved that if $E\rho^d = \infty$ then $\lambda_c(\rho) = 0$, while it follows easily from Theorem 3.3 that if we take ρ to be an a.s. positive constant, then $\lambda_c(\rho) > 0$. Combining these two facts, we can take a sequence $\rho_1, \rho_2, \ldots$ converging weakly to ρ such that $E\rho_k^d = \infty$ and ρ is an a.s. constant: in such a case, $\lambda_c(\rho_k) \not\to \lambda_c(\rho)$.

The next theorem gives a sufficient condition on the radii random variables to guarantee convergence of the critical density.

Theorem 3.7 *Let ρ_k and ρ be random variables such that for some $R > 0$ we have $0 \leq \rho \leq R$ and $0 \leq \rho_k \leq R$ a.s. for all $k \geq 1$. If $\rho_k \Rightarrow \rho$ then $\lambda_c(\rho_k) \to \lambda_c(\rho)$.*

Our strategy will be to approximate the radii ρ and ρ_k by radii which take only finitely many values. For ease of notation, we shall restrict ourselves to two dimensions. The proof for higher dimensions is similar.

Lemma 3.4 *Let $0 < r_1 < r_2 < \cdots < r_n < \infty$ and let ρ and ρ' be random variables taking values r_i with probability p_i and p'_i, respectively, for $i = 1, 2, \ldots, n$, where $\sum_{i=1}^{n} p_i = \sum_{i=1}^{n} p'_i = 1$. Suppose that there exist $1 \leq j < l \leq n$ such that for all $i \neq j$ or l and $i = 1, \ldots, n$, $p_i = p'_i$ and p_l and p'_l are both positive. Then,*

$$|\lambda_c(\rho) - \lambda_c(\rho')| \leq \frac{\lambda_c(r_1)}{\min\{p_l, p'_l\}} |p_j - p'_j|. \tag{3.70}$$

Proof Suppose first that $p_j > p'_j$. We shall use a coupling argument to prove that

$$\lambda_c(\rho) \geq \lambda_c(\rho'). \tag{3.71}$$

To see this, consider n independent Poisson processes $X_1, X_2, \ldots, X_n$ of densities $p_1\lambda, p_2\lambda, \ldots, p_{j-1}\lambda, p'_j\lambda, p_{j+1}\lambda, \ldots, p_n\lambda$, respectively. At each point of the process X_i we centre a ball of radius r_i. Now consider another independent Poisson process X' of density $\lambda(p_j - p'_j)$. Note that if at each point of this process X' we centre a ball of radius r_j then the superposition of the models $(X_1, r_1, p_1\lambda), (X_2, r_2, p_2\lambda), \ldots, (X_{j-1}, r_{j-1}, p_{j-1}\lambda), (X_j, r_j, p'_j\lambda), (X_{j+1}, r_{j+1}, p_{j+1}\lambda), \ldots, (X_n, r_n, p_n\lambda)$ and $(X', r_j, (p_j - p'_j)\lambda)$ is a Poisson Boolean model (X, ρ, λ) where X is the superposition of X' and X_i, $i = 1, \ldots, n$. If, instead, at the points of the process X' we centre a ball of radius r_l and then superpose all the models, we obtain a Poisson Boolean model (X, ρ', λ). Since $r_j < r_l$, the occupied region in (X, ρ, λ) will be contained in the occupied region in (X, ρ', λ). Hence the existence of an unbounded component in the model (X, ρ, λ) will imply the existence of an unbounded component in the model (X, ρ', λ) which implies inequality (3.71). We have explained this in detail because we shall be using this kind of coupling results very often later without going into the details of the proof.

Now choose $\lambda > \lambda_c(\rho')$. Consider the models $(X_i, r_i, \lambda l_i)$, for $i = 1, \ldots, l-1, l+1, \ldots, n$, where the l_i's are chosen such that

$$\frac{\lambda p'_i + \lambda l_i}{\lambda(1 + L)} = p_i, \quad i = 1, \ldots, l-1, l+1, \ldots, n, \tag{3.72}$$

for $L := l_1 + \cdots + l_{l-1} + l_{l+1} + \cdots + l_n$. The system of linear equations (3.72) can be explicitly solved to yield $l_i = (p_l)^{-1}(p_i p'_l - p_l p'_i) \geq 0$. Next, let $(\tilde{X}, \rho', \lambda)$ be a Boolean model independent of the models $(X_i, r_i, \lambda l_i)$ and consider the superposition of $(\tilde{X}, \rho', \lambda)$ and $(X_i, r_i, \lambda l_i)$, $i = 1, \ldots, l - 1$, $l + 1, \ldots n$ to obtain a model equivalent in law to $(X, \rho, \lambda(1 + L))$ where $X = X_1 * \cdots * X_{l-1} * X_{l+1} * \cdots * X_n * \tilde{X}$ and $*$ denotes superposition. (To see that the radius random variable in this superposition is ρ, just use (3.72).) Since $\lambda > \lambda_c(\rho')$, the model $(\tilde{X}, \rho', \lambda)$ is supercritical and hence the superposition is certainly supercritical. Thus

$$\lambda(1 + L) > \lambda_c(\rho).$$

The above inequality holds for all $\lambda > \lambda_c(\rho')$, so we have

$$\lambda_c(\rho')(1 + L) \geq \lambda_c(\rho).$$

We have from (3.72) and some elementary calculations $L = (p_l)^{-1}(p_j - p'_j)$. The result now follows since $\lambda_c(\rho') \leq \lambda_c(r_1)$. □

For the case $p_j < p'_j$, we just reverse the roles of ρ and ρ'. A similar argument as above yields (3.70). □

Lemma 3.5 *Let $0 < r_1 < \cdots < r_n$, and let ρ be a random variable taking values r_i with probability p_i for $i = 1, 2, \ldots, n$, where $\sum_{i=1}^{n} p_i = 1$. Suppose that $p_n > 0$. For all $k = 1, 2, \ldots$ define the random variables ρ_k taking values r_i with probability $p_{k,i}$ for all $i = 1, \ldots, n$, where $\sum_{i=1}^{n} p_{k,i} = 1$ for all k. If $p_{k,i} \to p_i$ for all i when $k \to \infty$, then $\lambda_c(\rho_k) \to \lambda_c(\rho)$.*

Proof We have assumed that $p_n > 0$ so we can pick $0 < \delta < p_n$. Take k_0 so large that $\sum_{i=1}^{n-1} |p_{k,i} - p_i| < \frac{1}{2}\delta$ for all $k \geq k_0$. Then, of course, we have $p_{k,n} > \frac{1}{2}\delta$ for all $k \geq k_0$. For $l = 1, \ldots, n - 1$ and $k \geq k_0$ let $\xi_k^{(l)}$ be the random variable defined by

$$P(\xi_k^{(l)} = r_i) = \begin{cases} p_{k,i} & \text{for } i = 1, \ldots, l, \\ p_i & \text{for } i = l+1, \ldots, n-1, \\ p_n + \sum_{i=1}^{l}(p_i - p_{k,i}) & \text{for } i = n. \end{cases}$$

Clearly, $\xi_k^{(n-1)}$ has the same distribution as ρ_k and we define $\xi_k^{(0)} := \rho$.
According to Lemma 3.4, for $l = 1, \ldots, n - 1$, we have

$$|\lambda_c(\xi_k^{(l)}) - \lambda_c(\xi_k^{(l-1)})| \leq 2\delta^{-1}\lambda_c(r_1)|p_l - p_{k,l}|.$$

Adding the previous inequalities over all l, and using the triangle inequality, we obtain

$$|\lambda_c(\rho_k) - \lambda_c(\rho)| \leq 2\delta^{-1}\lambda_c(r_1) \sum_{l=1}^{n-1} |p_l - p_{k,l}|,$$

for all $k \geq k_0$. This proves the lemma. □

Next we drop the assumption that p_n should be positive:

Lemma 3.6 *Let $0 < r_1 < \cdots < r_n$, and, for $k \geq 1$, let ρ and ρ_k be random variables taking values r_i with probability p_i and $p_{k,i}$ for $i = 1, 2, \ldots, n$, where $\sum_{i=1}^{n} p_i = \sum_{i=1}^{n} p_{k,i} = 1$. If $p_{k,i} \to p_i$ as $k \to \infty$ for all $1 \leq i \leq n$ then $\lambda_c(\rho_k) \to \lambda_c(\rho)$ as $k \to \infty$.*

Proof   In view of Lemma 3.5, we need to prove this lemma for the case when there exists $1 \leq m \leq n - 1$ such that

$$p_m > 0 \text{ and } p_{m+1} = \ldots = p_n = 0. \tag{3.73}$$

First we show that it suffices to prove the lemma for the case $m = n - 1$. Let the random variables ξ_k' and ξ_k'' be defined by

$$P(\xi_k' = r_i) = \begin{cases} p_{k,i} & \text{for } i = 1, \ldots, m, \\ 0 & \text{for } i = m + 1, \ldots, n - 1, \\ \sum_{i=m+1}^{n}(p_i - p_{k,i}) & \text{for } i = n, \end{cases}$$

and

$$P(\xi_k'' = r_i) = \begin{cases} p_{k,i} & \text{for } i = 1, \ldots, m, \\ \sum_{i=m+1}^{n}(p_i - p_{k,i}) & \text{for } i = m + 1, \\ 0 & \text{for } i = m + 1, \ldots, n, \end{cases}$$

then we clearly have

$$\lambda_c(\xi_k') \leq \lambda_c(\rho_k) \leq \lambda_c(\xi_k'').$$

So it suffices to show that $\lambda_c(\rho_k)$ converges to $\lambda_c(\rho)$ when the ρ_k's take at most one value larger than r_m with positive probability. Thus we henceforth assume that $m = n - 1$; i.e. $p_{n-1} > 0$ and $p_n = 0$.

Next let ρ_k' be a random variable taking values $r_1, \ldots, r_n$ with probabilities $p_1, p_2, \ldots, p_{n-2}, p_{k,n-1}', p_{k,n}$, respectively, where $p_{k,n-1}' := p_{n-1} - p_{k,n}$. For k large enough, since $p_{n-1} > 0$ and $p_{k,n} \to 0$ as $k \to \infty$, so $p_{k,n-1}' \geq 0$. We shall now prove

$$\lim_{k \to \infty} \lambda_c(\rho_k') = \lambda_c(\rho). \tag{3.74}$$

We observe from our choice of ρ_k' that

$$\lambda_c(\rho_k') \leq \lambda_c(\rho).$$

So to prove (3.74) we need to show that $\liminf_{k \to \infty} \lambda_c(\rho_k') \geq \lambda_c(\rho)$. Suppose there exists a λ such that $\liminf_{k \to \infty} \lambda_c(\rho_k') < \lambda < \lambda_c(\rho)$. Since $\lambda < \lambda_c(\rho)$, for κ_0 as in Lemma 3.3 and using Theorem 3.5, we can find an N such that the crossing probability in (X, ρ, λ) satisfies

$$\sigma((N, 3N), \lambda, 1) < \tfrac{1}{2}\kappa_0. \tag{3.75}$$

Now we construct independent Poisson Boolean models $(X_i, r_i, \lambda l_{k,i})$ for $i = 1, 2, \ldots, n-2, n$ and another independent Poisson Boolean model (X', ρ, λ) so as to yield the model $(X, \rho_k, \lambda(1 + L_k))$ when all the models are superposed, where $X = X' * X_1 * \cdots * X_{n-2} * X_n$ and $L_k = l_{k,1} + \cdots + l_{k,n-2} + l_{k,n}$. For this, we choose $l_{k,1}, \ldots, l_{k,n-2}, l_n$ to satisfy the following relations:

$$\frac{p_i + l_{k,i}}{1 + L_k} = p_i \quad \text{for } i = 1, \ldots, n-2, \tag{3.76}$$

and

$$\frac{l_{k,n}}{1 + L_k} = p_{k,n}. \tag{3.77}$$

The system of linear equations (3.76) and (3.77) can be explicitly solved to yield

$$l_{k,i} = \left(\frac{p_{n-1} - p'_{k,n-1}}{p'_{k,n-1}} \right) p_i \geq 0 \quad \text{for } i = 1, \ldots, n-2,$$

and

$$l_{k,n} = \frac{p_{n-1}}{p'_{k,n-1}} p_{k,n} \geq 0.$$

Clearly, for every $i = 1, \ldots, n-2$ and $i = n$, $l_{k,i} \to 0$ when $k \to \infty$. Thus, we can choose k large enough such that for all $i = 1, \ldots, n-2$ and $i = n$, we have

$$P_{\lambda l_{k,i}}(X_i([-R, N+R] \times [-R, 3N+R]) \geq 1) < \kappa_0/2n, \tag{3.78}$$

where κ_0 is as chosen before.

The superposition of the Poisson Boolean models $(X_i, r_i, \lambda l_{k,i})$ for all $i = 1, \ldots, n-2, n$ and (X', ρ, λ) is equivalent in law to the Poisson Boolean model $(X, \rho'_k, \lambda(1 + L_k))$. For k large enough, (3.75) and (3.78) imply that $\sigma((N, 3N), \lambda(1 + L_k), 1) < \kappa_0$, and thus it follows from Lemma 3.3, that the superposed model is subcritical. However, by the choice of λ, (X, ρ'_k, λ) is supercritical, hence so is $(X, \rho'_k, \lambda(1 + L_k))$ which is the desired contradiction.

Finally, to complete the proof of the lemma, we construct $\xi_k^{(l)}$ as in the previous lemma, where $\xi_k^{(n-1)}$ has the same distribution as ρ_k and $\xi_k^{(0)} = \rho'_k$.

This method shows that

$$|\lambda_c(\rho_k) - \lambda_c(\rho_k')| \leq 2(p_{n-1})^{-1}\lambda_c(r_1) \sum_{i=1}^{n-2} |p_{k,i} - p_i|,$$

and the lemma follows.   □

Proof of Theorem 3.7   First we suppose that the supports of both ρ and ρ_k, $k = 1, 2, \ldots$ are contained in an interval $[a, R]$, where $a > 0$. The distribution function of ρ is denoted by F, and the distribution function of ρ_k by F_k. We can assume that both a and R are continuity points of F. Take a sequence $\{\pi_n\}$ of partitions of $[a, R]$, which we write as $\pi_n = \{a = \gamma_0^n < \gamma_1^n < \cdots < \gamma_{k_n}^n = R\}$. The partitions are chosen in such a way that π_{n+1} refines π_n, all points γ_i^n are continuity points of F and such that $|\pi_n| := \max_{1 \leq i \leq k_n} \{\gamma_i^n - \gamma_{i-1}^n\} \to 0$ when $n \to \infty$. Now define, for all $n \geq 1$, the random variables $\rho^{(n)}$ and $\rho_{(n)}$ by the requirement that if $\rho \in (\gamma_{i-1}^n, \gamma_i^n]$, then $\rho^{(n)} = \gamma_i^n$ and $\rho_{(n)} = \gamma_{i-1}^n$. It follows from a simple coupling argument that $\lambda_c(\rho^{(n)}) \leq \lambda_c(\rho) \leq \lambda_c(\rho_{(n)}) \leq \lambda_c(a)$. Also, it is easy to see that $\lambda_c(\rho^{(n)})$ is increasing and $\lambda_c(\rho_{(n)})$ is decreasing in n. Now write

$$\alpha_n := \max_{1 \leq i \leq k_n} \frac{\gamma_i^n}{\gamma_{i-1}^n} \leq 1 + \frac{|\pi_n|}{a},$$

which tends to 1 when $n \to \infty$. Hence $\rho^{(n)} \leq \alpha_n \rho_{(n)}$, which implies that $\lambda_c(\rho^{(n)}) \geq \lambda_c(\alpha_n \rho_{(n)}) = \alpha_n^{-2}\lambda_c(\rho_{(n)})$ and thus

$$\lambda_c(\rho^{(n)}) \leq \lambda_c(\rho) \leq \alpha_n^2 \lambda_c(\rho^{(n)}).$$

We can now write

$$\lambda_c(\rho) - \lambda_c(\rho^{(n)}) \leq \left[\left(1 + \frac{|\pi_n|}{a}\right)^2 - 1\right]\lambda_c(\rho^{(n)})$$

$$\leq \left[\left(1 + \frac{|\pi_n|}{a}\right)^2 - 1\right]\lambda_c(a) =: \beta_n \text{ (say)}. \qquad (3.79)$$

The previous calculation can also be done for ρ_k instead of ρ and we obtain, in the obvious notation

$$\lambda_c(\rho_k) - \lambda_c(\rho_k^{(n)}) \leq \beta_n. \qquad (3.80)$$

Now given any $\epsilon > 0$, take n so large that $\beta_n < \epsilon$. Observe that $\rho^{(n)}$ takes the value γ_i^n with probability $F(\gamma_i^n) - F(\gamma_{i-1}^n)$ and $\rho_k^{(n)}$ takes the value γ_i^n with probability $F_k(\gamma_i^n) - F_k(\gamma_{i-1}^n)$. Hence by the choice of the partitions, the fact that $\rho_k \Rightarrow \rho$ and Lemma 3.6, we see that $|\lambda_c(\rho^{(n)}) - \lambda_c(\rho_k^{(n)})| < \epsilon$ for

k sufficiently large. Together with (3.79) and (3.80) this proves the theorem in this case.

Next we drop the assumption that the supports are bounded from below by some positive number. Let $\delta > 0$ be a continuity point of F and let $\eta > 0$ be such that $P_{(\lambda,\rho)}(\rho > \delta) > \eta$. Since $\rho_k \Rightarrow \rho$, we have $P_{(\lambda,\rho_k)}(\rho_k > \delta) > \eta$ for k sufficiently large. Certainly, if $(X', \delta, \eta\lambda)$ is supercritical, so is (X, ρ_k, λ) and it follows that if $\eta\lambda > \lambda_c(\delta)$ then $\lambda > \lambda_c(\rho_k)$, or

$$\lambda_c(\rho_k) \leq \frac{1}{\eta}\lambda_c(\delta). \tag{3.81}$$

Now let $\epsilon > 0$ and choose a to be a continuity point of F such that $F(a) < \epsilon$, and choose k_0 so large that $F_k(a) < \epsilon$ for all $k \geq k_0$. Let ρ^a be a random variable with distribution equal to the conditional distribution of ρ, given that $\rho \geq a$. Similarly, let ρ_a be a random variable with distribution equal to the conditional distribution of ρ given $\rho < a$. Then we have $\lambda_c(\rho^a) \leq \lambda_c(\rho)$.

Consider the model (X_1, ρ^a, λ) and $(X_2, \rho_a, \lambda l)$, where l is chosen such that $l(1 + l)^{-1} = P_{(\lambda,\rho)}(\rho \leq a)$. This means that

$$l = \frac{F(a)}{1 - F(a)}. \tag{3.82}$$

The superposition of the two models is equivalent in law to a process $(X, \rho, \lambda(1 + l))$. The following formula is obtained as in (3.52):

$$|\lambda_c(\rho_k) - \lambda_c(\rho_k^a)| \leq \frac{\epsilon}{1 - \epsilon}\lambda_c(\rho_k^a) \leq \frac{\epsilon}{\eta(1 - \epsilon)}\lambda_c(\delta). \tag{3.83}$$

When $\rho_k \Rightarrow \rho$, then $\rho_k^a \Rightarrow \rho^a$ and from the case already proved we conclude that

$$|\lambda_c(\rho_k^a) - \lambda(\rho^a)| < \epsilon \tag{3.84}$$

for k large enough. The result now follows from (3.52), (3.83) and (3.84). $\quad\square$

The obvious question arises as to whether or not the percolation probabilities also converge when the radii are uniformly bounded and $\rho_k \Rightarrow \rho$. Note that pointwise convergence of the percolation function does not imply convergence of the critical densities.

Theorem 3.8 *Let ρ_k and ρ be random variables such that for some $R > 0$ we have $0 < \rho \leq R$ and $0 < \rho_k \leq R$ a.s. for all $k \geq 1$. If $\rho_k \Rightarrow \rho$, then $\theta_{\rho_k}(\lambda) \to \theta_\rho(\lambda)$ for all $\lambda \neq \lambda_c(\rho)$.*

In the proof of this theorem we shall need the fact that $\theta_\rho(\lambda)$ is, for fixed ρ, a continuous function of λ, for $\lambda \neq \lambda_c(\rho)$.

Theorem 3.9   *In a Poisson Boolean model* (X, ρ, λ), *the percolation function* θ_ρ *is a continuous function of* λ *for all* $\lambda \neq \lambda_c(\rho)$.

Proof of Theorem 3.9   First we show that θ is continuous from the right. The event $\{d(W) = \infty\}$ is the decreasing limit of the events $E_n = \{0 \overset{o}{\rightsquigarrow} \partial(B_n)\}$ where $B_n = [-n, n]^d$. We claim that $P_\lambda(E_n)$ is continuous in λ. To see this, first note that it follows from Lemma 3.1 that if we let $\delta \to 0$, the probability that in the Boolean model (X, ρ, δ) there is a ball intersecting B_n tends to zero. Thus if we couple two Boolean models with densities λ and $\lambda + \delta$ respectively on the same probability space as usual, the probability that $\{0 \overset{o}{\rightsquigarrow} \partial(B_n)\}$ in one model but not in the other tends to 0 when $\delta \to 0$. Thus we obtain that θ is a decreasing limit of a sequence of non-decreasing and continuous functions. This implies that it is continuous from the right.

Continuity from the left requires more work. Fix $\lambda_0 > \lambda_c(\rho)$ and take any $\lambda \in (\lambda_c(\rho), \lambda_0)$. Let $\alpha := (\lambda/\lambda_0)^{1/d}$ and scale (X, ρ, λ) by α to obtain the model $(\alpha X, \alpha\rho, \lambda_0)$. It follows immediately from the scaling that $\theta_{\alpha\rho}(\lambda_0) = \theta_\rho(\lambda)$. Now define

$$\psi(\alpha) := \theta_{\alpha\rho}(\lambda_0) = \theta_\rho(\lambda_0\alpha^d).$$

It is enough to prove that ψ is continuous from the left at 1, and this is what we shall do.

We couple all processes $(X, \alpha\rho, \lambda_0)$, $0 \leq \alpha \leq 1$ in the (by now) obvious way and we denote by W_α the occupied component of the origin in $(X, \alpha\rho, \lambda_0)$. In this coupling we clearly have $W_{\alpha_1} \subseteq W_{\alpha_2}$ whenever $\alpha_1 \leq \alpha_2$. It suffices to prove

$$P(d(W_1) = \infty, d(W_\alpha) < \infty \text{ for all } \alpha < 1) = 0. \tag{3.85}$$

In order to prove (3.85) we need to show that for almost all configurations for which $d(W_1) = \infty$, there exists a $\beta < 1$ (depending on the configuration!) such that also $d(W_\beta) = \infty$. First we claim that in any Boolean model (X, ρ, λ), the probability that two balls have exactly one point in common is zero. To see this, let ρ_0 be the radius of the ball centred at the origin and first condition on $\rho_0 = s$, say. Then the probability that the ball centred at the point x has exactly one point in common with the ball centred at the origin is equal to $P(\rho = |x| - s)$. According to Proposition 1.3, the number of such balls is Poisson distributed with parameter $\lambda \int_{\mathbb{R}^d} P(\rho = |x| - s)dx$. The integrand, however, is almost surely equal to zero and the claim now follows by integrating over s.

Now suppose that $d(W_1) = \infty$ and let $\alpha < 1$ be as defined previously. For this choice of α we have $\theta_{\alpha\rho}(\lambda_0) = \theta_\rho(\lambda) > 0$. We know from Theorem 3.6

that in $(X, \alpha\rho, \lambda_0)$ there is almost surely exactly one unbounded occupied component U_α, say, which must be contained in W_1 by the coupling. If the origin is contained in U_α we are done, so suppose it is not. In that case there exists a.s. a sequence $(0 = x_0, x_1, \ldots, x_{n-1}, x_n)$ of points of the point process such that

(i) $x_n \in U_\alpha$.
(ii) $d(x_i, x_{i+1}) < r_i + r_{i+1}$, where r_i is the radius of the ball centred at x_i. (The *strict* inequality here follows from the claim above.)

Now choose $\alpha' < 1$ such that $d(x_i, x_{i+1}) \leq \alpha'(r_i + r_{i+1})$ for all $i = 0, \ldots, n - 1$ and let $\beta := \max\{\alpha, \alpha'\}$. It is clear from the construction that $d(W_\beta) = \infty$ and the proof is complete. $\qquad\square$

REMARK: In Chapter 4 we shall prove that, in two dimensions, the percolation function θ_ρ is also continuous at criticality when the balls are bounded. The necessary machinery for this will be developed in Chapter 4.

For the proof of Theorem 3.8, we first note from Theorem 3.7 that if $\lambda < \lambda_c(\rho)$, then for n large $\lambda < \lambda_c(\rho_n)$. So $\theta_\rho(\lambda) = \theta_{\rho_n}(\lambda) = 0$ for n large and $\lambda < \lambda_c$. Thus we need only consider $\lambda > \lambda_c(\rho)$. As in the proof of Theorem 3.7, we shall approximate the radius random variable by random variables which take only finitely many values. The approximation techniques used to prove this theorem are similar to those used to prove Theorem 3.7.

Lemma 3.7 *Let $0 < r_1 < r_2 < \cdots < r_n < \infty$ and let ρ and ρ' be random variables taking values r_i with probability p_i and p_i', respectively, for $i = 1, \ldots, n$, where $\sum_{i=1}^n p_i = \sum_{i=1}^n p_i' = 1$. Suppose that there exist $1 \leq j < l \leq n$ such that $p_i = p_i'$ for all $i \neq j, l$ and where p_l and p_l' are both positive. Then,*

$$\theta_\rho\left(\frac{\lambda}{1 + (p_l')^{-1}|p_j - p_j'|}\right) \leq \theta_{\rho'}(\lambda) \leq \theta_\rho(\lambda(1 + (p_l)^{-1}|p_j - p_j'|)).$$

Proof Suppose first that $p_j > p_j'$. By a coupling argument as before we obtain

$$\theta_\rho(\lambda) \leq \theta_{\rho'}(\lambda). \tag{3.86}$$

As in the proof of Lemma 3.4, we consider the models $(X, r_i, \lambda l_i)$, for $i = 1, \ldots, l-1, l+1, \ldots, n$, where the l_i's are chosen as in (3.72). Next, consider the superposition of $(\tilde{X}, \rho', \lambda)$ and $(X_i, r_i, \lambda l_i), i = 1, \ldots, l-1, l+1, \ldots, n$ to obtain a model equivalent in law to $(X, \rho, \lambda(1 + L))$ where $L = l_1 + \cdots + l_{l-1} + l_{l+1} + \cdots + l_n$. By a coupling argument we obtain

$$\theta_\rho(\lambda(1 + L)) \geq \theta_{\rho'}(\lambda). \tag{3.87}$$

Now $L = (p_l)^{-1}(p_j - p'_j)$ and thus we have

$$\theta_\rho(\lambda(1 + L)) \geq \theta_{\rho'}(\lambda) \geq \theta_\rho(\lambda) \geq \theta_\rho(\lambda/(1 + L')),$$

where $L' = (p'_l)^{-1}|p_j - p'_j|$ and the last inequality follows by the non-decreasingness of the percolation function.

In case $p_j < p'_j$ we can repeat the argument starting with a density $\lambda' = \lambda/(1 + L')$ with the roles of ρ and ρ' interchanged.   □

Lemma 3.8 *Let $0 < r_1 < \cdots < r_n$, and let ρ be a random variable taking values r_i with probability p_i for $i = 1, \ldots, n$, where $\sum_{i=1}^n p_i = 1$. Suppose that $p_n > 0$. For all $k = 1, 2, \ldots$, define the random variables ρ_k taking values r_i with probability $p_{k,i}$, for all $i = 1, \ldots, n$, where $\sum_{i=1}^n p_{k,i} = 1$ for all k. If $p_{k,i} \to p_i$ as $k \to \infty$ for all i, then $\theta_{\rho_k}(\lambda) \to \theta_\rho(\lambda)$ as $k \to \infty$ for all $\lambda > \lambda_c(\rho)$.*

Proof The proof of this lemma is similar to that of Lemma 3.5. We first choose $0 < \delta < p_n$ and take k_0 so large that $\sum_{i=1}^{n-1} |p_{k,i} - p_i| < \frac{1}{2}\delta$, for all $k \geq k_0$. Then, of course, we have $p_{k,n} > \frac{1}{2}\delta$, for all $k \geq k_0$. So, by using Lemma 3.7, we obtain,

$$\theta_\rho\left(\lambda \prod_{i=1}^{n-1}\left(1 - \frac{2|p_{k,i} - p_i|}{\delta}\right)\right) \leq \theta_{\rho_k}(\lambda) \leq \theta_\rho\left(\lambda \prod_{i=1}^{n-1}\left(1 + \frac{2|p_{k,i} - p_i|}{\delta}\right)\right).$$

Thus,

$$|\theta_{\rho_k}(\lambda) - \theta_\rho(\lambda)|$$
$$\leq \theta_\rho\left(\lambda \prod_{i=1}^{n-1}\left(1 + \frac{2|p_{k,i} - p_i|}{\delta}\right)\right) - \theta_\rho\left(\lambda \prod_{i=1}^{n-1}\left(1 - \frac{2|p_{k,i} - p_i|}{\delta}\right)\right).$$

Now, by continuity of $\theta_\rho(\lambda)$ for $\lambda > \lambda_c(\rho)$ (Theorem 3.9), we have

$$\theta_\rho\left(\lambda \prod_{i=1}^{n-1}\left(1 + \frac{2|p_{k,i} - p_i|}{\delta}\right)\right) - \theta_\rho\left(\lambda \prod_{i=1}^{n-1}\left(1 - \frac{2|p_{k,i} - p_i|}{\delta}\right)\right) \to 0$$

as $k \to \infty$.   □

Lemma 3.9 *Let $0 < r_1 < \cdots < r_n$ and for $k \geq 1$, suppose ρ and ρ_k are random variables taking value r_i with probability p_i and $p_{k,i}$, respectively, for all $1 \leq i \leq n$, where $\sum_{i=1}^n p_i = \sum_{i=1}^n p_{k,i} = 1$. If $p_{k,i} \to p_i$ as $k \to \infty$, for all $1 \leq i \leq n$, then $\lambda_c(\rho_k) \to \lambda_c(\rho)$ as $k \to \infty$.*

Proof We need to prove this lemma for the case when there exists $1 \le m \le n - 1$ such that

$$p_m > 0 \text{ and } p_{m+1} = \cdots = p_n = 0.$$

The same argument as in the proof of Lemma 3.6 shows that we may assume that $m = n - 1$; i.e. $p_{n-1} > 0$ and $p_n = 0$. Using the same idea as before, it is enough to prove the result in case $p_{k,i} = p_i$ for all $i = 1, 2, \ldots, n - 2$ for each $k \ge 1$. Also we may assume that $p_{k,n}$ decreases to zero as $k \to \infty$.

Now, let $B_M = [-M, M]^d$ and $\partial(B_M)$ be the boundary of B_M. Then, for every $k \ge 1$, it is not hard to see that

$$P_{(\lambda, \rho_k)}(0 \overset{\circ}{\rightsquigarrow} \partial(B_M)) \downarrow \theta_{\rho_k}(\lambda) \text{ as } M \to \infty.$$

Similarly,

$$P_{(\lambda, \rho)}(0 \overset{\circ}{\rightsquigarrow} \partial(B_M)) \downarrow \theta_{\rho}(\lambda) \text{ as } M \to \infty.$$

Fix an $M \ge 1$. We claim that

$$\lim_{k \to \infty} P_{(\lambda, \rho_k)}(0 \overset{\circ}{\rightsquigarrow} \partial(B_M)) = P_{(\lambda, \rho)}(0 \overset{\circ}{\rightsquigarrow} \partial(B_M)). \tag{3.88}$$

Clearly, for each $k \ge 1$, we have

$$P_{(\lambda, \rho_k)}(0 \overset{\circ}{\rightsquigarrow} \partial(B_M)) \ge P_{(\lambda, \rho_{k+1})}(0 \overset{\circ}{\rightsquigarrow} \partial(B_M)) \ge P_{(\lambda, \rho)}(0 \overset{\circ}{\rightsquigarrow} \partial(B_M)).$$

Hence

$$\lim_{k \to \infty} P_{(\lambda, \rho_k)}(0 \overset{\circ}{\rightsquigarrow} \partial(B_M)) \ge P_{(\lambda, \rho)}(0 \overset{\circ}{\rightsquigarrow} \partial(B_M)). \tag{3.89}$$

Given $\epsilon > 0$ we choose k large such that $1 - \exp(-\lambda(2M)^d p_{k,n}) < \epsilon$. Now we consider n independent Poisson processes $X_1, X_2, \ldots, X_n$ with densities $\lambda p_1, \lambda p_2, \ldots, \lambda p_{n-2}, \lambda p_{k,n-1}, \lambda p_{k,n}$ respectively. At each point of X_i, $1 \le i \le n - 1$, we centre a ball of radius r_i. For the n-th process X_n we distinguish two cases: (i) at each point of the process X_n we centre a ball of radius r_n and (ii) at each point of the process X_n we centre a ball of radius r_{n-1}. In case (i) we obtain the Poisson Boolean model (X, ρ_k, λ), while in case (ii) we obtain the Poisson Boolean model (X, ρ, λ). Thus by this coupling, we obtain,

$$P_{(\lambda, \rho_k)}(0 \overset{\circ}{\rightsquigarrow} \partial(B_M)) - P_{(\lambda, \rho)}(0 \overset{\circ}{\rightsquigarrow} \partial(B_M))$$
$$\le P(X_n([-M, M]^d) \ge 1)$$
$$= 1 - \exp(-\lambda(2M)^d p_{k,n})$$
$$< \epsilon.$$

This proves (3.88).

Now consider the double sequence $\{P_{(\lambda,\rho_k)}(0 \overset{o}{\leadsto} \partial(B_M))\}$ in k and M. Note that the sequence is decreasing in both M and k. Hence both the iterated limits exist and are equal. Hence,

$$
\begin{aligned}
\lim_{k \to \infty} \theta_{\rho_k}(\lambda) &= \lim_{k \to \infty} \lim_{M \to \infty} P_{(\lambda,\rho_k)}(0 \overset{o}{\leadsto} \partial(B_M)) \\
&= \lim_{M \to \infty} \lim_{k \to \infty} P_{(\lambda,\rho_k)}(0 \overset{o}{\leadsto} \partial(B_M)) \\
&= \lim_{M \to \infty} P_{(\lambda,\rho)}(0 \overset{o}{\leadsto} \partial(B_M)) \\
&= \theta_\rho(\lambda).
\end{aligned}
\tag{3.90}
$$

$\square$

Proof of Theorem 3.8   First we assume that for some $a > 0$,

$$a > 0 \text{ such that } a < \rho, \rho_k \le R \text{ for all } k \ge 1. \tag{3.91}$$

Our strategy is to approximate the random variables ρ and ρ_k by random variables which take only finitely many values. Let the distribution functions of ρ and ρ_k be denoted by F and F_k respectively. We can assume that both a and R are continuity points of F. Take a sequence $\{\pi_n\}$ of partitions of $[a, R]$, which we write as $\pi_n = \{a = \gamma_0^n < \gamma_1^n < \cdots < \gamma_{k_n}^n = R\}$. We choose the partitions in such a way that π_{n+1} is a refinement of π_n. Also assume that all points γ_i^n are continuity points of F and $|\pi_n| := \max_{1 \le i \le k_n} \{\gamma_i^n - \gamma_{i-1}^n\} \to 0$, as $n \to \infty$. Now define, for all $n \ge 1$, the random variables $\rho^{(n)}$ and $\rho_{(n)}$ by the requirement that if $\rho \in (\gamma_{i-1}^n, \gamma_i^n]$, then $\rho^{(n)} = \gamma_i^n$ and $\rho_{(n)} = \gamma_{i-1}^n$. It follows from a simple coupling argument that $\theta_{\rho_{(n)}}(\lambda) \le \theta_\rho(\lambda) \le \theta_{\rho^{(n)}}(\lambda)$. Now for each $k \ge 1$, define the random variables $\rho_{k,(n)}$ and $\rho_k^{(n)}$ as follows: if $\rho_k \in (\gamma_{i-1}^n, \gamma_i^n]$, then $\rho_{k,(n)} = \gamma_{i-1}^n$ and $\rho_k^{(n)} = \gamma_i^n$. Clearly, for each $n \ge 1$ and $k \ge 1$, we have $\theta_{\rho_{k,(n)}}(\lambda) \le \theta_{\rho_k}(\lambda) \le \theta^{\rho_k^{(n)}}(\lambda)$.

Now, given $\epsilon > 0$, choose

$$\lambda_c(\rho) < \lambda_1 < \lambda < \lambda_2 \tag{3.92}$$

such that

$$\theta_\rho(\lambda_2) - \theta_\rho(\lambda_1) < \epsilon. \tag{3.93}$$

(Note that here we use the continuity of θ_ρ w.r.t. λ.) As before, let for each $n \ge 1$,

$$\alpha_n := \max_{1 \le i \le k_n} \frac{\gamma_i^n}{\gamma_{i-1}^n} \le 1 + \frac{|\pi_n|}{a},$$

which tends to 1 as $n \to \infty$. Note that $\rho^{(n)} \leq \alpha_n \rho_{(n)}$. Applying a change of scale to $(X, \rho_{(n)}, \lambda_0)$, we obtain the model $(\alpha_n X,, \alpha_n \rho_{(n)}, (\alpha_n)^{-d} \lambda_0)$. Since $\rho^{(n)} \leq \alpha_n \rho_{(n)}$, we have for any $\lambda_0 > 0$

$$\theta_{\rho_{(n)}}(\lambda_0) = \theta_{\alpha_n \rho_{(n)}}((\alpha_n)^{-d} \lambda_0) \geq \theta_{\rho^{(n)}}((\alpha_n)^{-d} \lambda_0). \tag{3.94}$$

Choose n large such that $(\alpha_n)^{-d} \lambda > \lambda_2$ and $(\alpha_n)^d \lambda < \lambda_1$. Now, by the choice of the partitions and the fact that $\rho_k \Rightarrow \rho$ we note that the random variables $\rho_k^{(n)}$ and $\rho^{(n)}$ satisfy the conditions of Lemma 3.9. Thus applying the lemma, we obtain that there exists K_1 such that for $k \geq K_1$

$$|\theta_{\rho_k^{(n)}}(\lambda) - \theta_{\rho^{(n)}}(\lambda)| < \epsilon. \tag{3.95}$$

For $k \geq K_1$, we may now write

$$\begin{aligned}
\theta_{\rho_k}(\lambda) - \theta_\rho(\lambda) &\leq \theta_{\rho_k^{(n)}}(\lambda) - \theta_\rho(\lambda) \\
&\leq \theta_{\rho^{(n)}}(\lambda) + \epsilon - \theta_\rho(\lambda) \text{ by (3.95)} \\
&\leq \theta_{\rho_{(n)}}(\alpha_n^d \lambda) + \epsilon - \theta_\rho(\lambda) \text{ by (3.94)} \\
&\leq \theta_\rho(\alpha_n^d \lambda) + \epsilon - \theta_\rho(\lambda) \\
&\leq \theta_\rho(\lambda_2) + \epsilon - \theta_\rho(\lambda) \\
&\leq 2\epsilon,
\end{aligned}$$

where the last inequality follows from (3.92) and (3.93).

Similarly, for fixed n we can choose K_2 so large that for $k \geq K_2$ we have

$$\theta_\rho(\lambda) - \theta_{\rho_k}(\lambda) \leq 2\epsilon.$$

This proves the theorem in case (3.91) holds.

Now let ρ have support $(0, R]$. Here again, given $\epsilon > 0$, choose $\lambda_1 < \lambda < \lambda_2$ such that (3.93) holds. Let a be a continuity point of F, the distribution function of ρ. Let ρ^a be a random variable with distribution equal to the conditional distribution of ρ, given that $\rho \geq a$. Similarly, let ρ_a be a random variable with distribution equal to the conditional distribution of ρ given $\rho < a$. Then we have $\theta_\rho(\lambda) \leq \theta_{\rho^a}(\lambda)$. Also define ρ_k^a as the random variable having distribution equal to the conditional distribution of ρ_k, given that $\rho_k \geq a$ and $\rho_{k,a}$ as the random variable having the distribution function equal to the conditional distribution of ρ_k, given $\rho_k < a$. Note that $\theta_{\rho_k}(\lambda) \leq \theta_{\rho_k^a}(\lambda)$.

For any $\lambda_0 > 0$, consider the models (X_1, ρ^a, λ_0) and $(X_2, \rho_a, \lambda_0 l)$, where l is such that $l(1 + l)^{-1} = F(a)$; i.e.,

$$l = \frac{F(a)}{1 - F(a)}. \tag{3.96}$$

The superposition of these two models is equivalent in law to the model $(X, \rho, \lambda_0(1 + l))$. Thus, we obtain

$$\theta_\rho(\lambda_0(1 + l)) \geq \theta_{\rho^a}(\lambda_0). \tag{3.97}$$

The same calculations may be carried out for ρ_k with $l_k = F_k(a)/(1 - F_k(a))$, giving

$$\theta_{\rho_k}(\lambda_0(1 + l_k)) \geq \theta_{\rho_k^a}(\lambda_0). \tag{3.98}$$

Now, as $a \to 0$, $F(a) \to 0$. Thus we may choose a small enough so that $\lambda(1 + l) < \lambda_2$ and $\lambda/(1 + l) > \lambda_1$. Now, for this a, we have $F_k(a) \to F(a)$ as $k \to \infty$. Choose K_3 large so that $\lambda(1 + l_k) < \lambda_2$ and $\lambda/(1 + l_k) > \lambda_1$ for $k \geq K_3$.

Now, the random variables ρ_k^a and ρ^a are bounded below by a. Also, $\rho_k^a \Rightarrow \rho^a$ as $\rho_k \Rightarrow \rho$ and a is a continuity point of F. Hence by the first part of the argument, we may choose K_4 large so that for $k \geq K_4$ we have

$$|\theta_{\rho_k^a}(\lambda) - \theta_{\rho^a}(\lambda)| < \epsilon \tag{3.99}$$

and

$$|\theta_{\rho_k^a}(\lambda_1) - \theta_{\rho^a}(\lambda_1)| < \epsilon. \tag{3.100}$$

Thus, we have from (3.93), (3.98) and (3.100) with $\lambda_0 = \lambda/(1 + l_k)$,

$$
\begin{aligned}
\theta_\rho(\lambda) - \theta_{\rho_k}(\lambda) &\leq \theta_\rho(\lambda) - \theta_{\rho_k^a}(\lambda/(1 + l_k)) \\
&\leq \theta_\rho(\lambda) - \theta_{\rho_k^a}(\lambda_1) \\
&\leq \theta_\rho(\lambda) - \theta_{\rho^a}(\lambda_1) + \epsilon \\
&\leq \theta_\rho(\lambda) - \theta_\rho(\lambda_1) + \epsilon \\
&\leq 2\epsilon,
\end{aligned}
$$

and using (3.93), (3.97) and (3.99) with $\lambda_0 = \lambda$,

$$
\begin{aligned}
\theta_{\rho_k}(\lambda) - \theta_\rho(\lambda) &\leq \theta_{\rho_k^a}(\lambda) - \theta_\rho(\lambda) \\
&\leq \theta_{\rho^a}(\lambda) - \theta_\rho(\lambda) + \epsilon \\
&\leq \theta_\rho(\lambda(1 + l)) - \theta_\rho(\lambda) + \epsilon \\
&\leq \theta_\rho(\lambda_2) - \theta_\rho(\lambda) + \epsilon \\
&\leq 2\epsilon.
\end{aligned}
$$

This completes the proof of the theorem. $\square$

3.9 Bounds on λ_c and asymptotics for the cluster size

Consider a Boolean model $(X, 1, \lambda)$ in two dimensions. We have the following explicit bounds for the critical density:

Theorem 3.10 *For a Poisson Boolean model $(X, 1, \lambda)$ on $\mathbb{R}^2$ we have,*

$$0.174 < \lambda_c < 0.843.$$

Proof First we show that $\lambda_c > 0.174$ and for this we employ a multi-type branching process argument as in Theorem 3.2. The types we consider now are distributed over all real numbers in $(0, 2)$, unlike in the earlier cases when the types assumed non-negative integer values.

Let $x_1, x_2, \ldots$ be the points of the Poisson process X of density λ and fix x_1 to be the initial member of the 0-th generation of the branching process. We take another Poisson process X_1 of density λ, independent of X and let $x_{1,1}, x_{1,2}, \ldots, x_{1,n_1}$ be all the points of X_1 which lie in the ball $S(x_1, 2) = \{y : |y - x_1| \le 2\}$. The children of x_1 in this branching process are these points $x_{1,1}, x_{1,2}, \ldots, x_{1,n_1}$.

Let $x_{k,1}, x_{k,2}, \ldots, x_{k,n_k}$ be the members of the k-th generation of the branching process. To obtain the children of $x_{k,i}$, we consider a Poisson point process $X_{k+1,i}$ of density λ on $\mathbb{R}^2$, where $X_{k+1,i}$ is independent of all the processes described as yet. The children of $x_{k,i}$ are those points of the process $X_{k+1,i}$ which fall in the region $S(x_{k,i}, 2) \backslash S(x_{k-1,j}, 2)$, where $x_{k-1,j}$ is the parent of $x_{k,i}$. The type of a child $x_{k+1,l}$ of $x_{k,i}$ is $t := |x_{k,i} - x_{k+1,l}| \in (0, 2)$. Clearly, the distribution of the number and types of children of $x_{k,i}$ depend only on $x_{k,i}$ and its type. Indeed, the distribution of the number of children of $x_{k,i}$ whose types lie in $(a, b), 0 \le a < b \le 2$ depends only on the area of the region

$$(S(x_{k,i}, 2) \backslash S(x_{k-1,j}, 2)) \cap \{y : |y - x_{k,i}| \in (a, b)\},$$

and this area depends on $x_{k-1,j}$ only through the distance $|x_{k,i} - x_{k-1,j}|$, which is precisely the type of $x_{k,i}$. Also, the distribution of the number and types of children of an individual $x_{k,i}$ does not depend on its generation k.

Given that $x_{k,i}$ is of type u, i.e. $|x_{k,i} - x_{k-1,j}| = u$, let $g(v|u)$ be the length of the curve given by $(S(x_{k,i}, 2) \backslash S(x_{k-1,j}, 2)) \cap \{y : |y - x_{k,i}| = v\}$. A precise expression for $g(v|u)$ follows from an elementary trigonometric calculation, which yields

$$g(v|u) = \begin{cases} 2v \cos^{-1} \dfrac{4 - u^2 - v^2}{2uv} & \text{if } 2 - u < v < 2 \\ 0 & \text{if } 0 < v \le 2 - u. \end{cases} \quad (3.101)$$

Recalling our earlier discussion on the independence properties of the off-spring distribution, we easily see that the expected number of children whose types lie in (a, b) of an individual whose type is u is given by $\int_a^b \lambda g(v|u)dv$. Moreover, given that an individual is of type u, the expected total number of grandchildren of this individual whose types lie in (a, b) is given by

$$\int_0^2 \left(\int_a^b \lambda^2 g(v|w) \, dv \right) g(w|u) \, dw. \tag{3.102}$$

In other words, if we let

$$g_1(v|u) := \int_0^2 g(v|w) g(w|u) dw,$$

the integral in (3.102) reduces to

$$\lambda^2 \int_a^b g_1(v|u) \, dv.$$

Thus defining recursively,

$$g_n(v|u) := \int_0^2 g_{n-1}(v|w) g(w|u) \, dw,$$

we easily see that the expected number of members of the n-th generation having types in (a, b) coming from a particular individual of type u as an ancestor n generations previously is given by

$$\lambda^n \int_a^b g_n(v|u) \, dv.$$

Hence the expected total number of individuals in the branching process if we start off with an individual of type u is

$$\sum_{n=1}^\infty \lambda^n \int_0^2 g_n(v|u) \, dv. \tag{3.103}$$

To show that (3.103) converges for sufficiently small λ, we use an estimate based on the theory of Hilbert–Schmidt operators (see Dunford and Schwartz 1958, chap. XI, sec. 6). For all complex-valued, square integrable functions f defined on the interval $(0, 2)$, consider the linear operator T_f defined by

$$T_f(u) = \int_0^2 f(v) g(v|u) dv.$$

It is easy to see that

$$\int_0^2 g_n(v|u) dv = T_1^n(u)$$

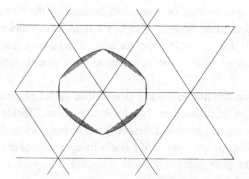

Figure 3.2. The triangular lattice with the 'flower' as described in the text.

where $1(v) \equiv 1$ for all $v \in (0, 2)$. Thus to show that (3.103) converges for sufficiently small λ we need to know that

$$\sum_{n=1}^{\infty} \lambda^n T_1^n(u) \qquad (3.104)$$

converges for sufficiently small λ. This is indeed true for $\lambda < \|T\|^{-1}$, where $\|T\|$ denotes the usual operator norm of T, i.e. $\|T\| = \sup\{\|T_f\|_2 : f$ square integrable, complex-valued functions on $(0, 2)$ with $\|f\|_2 \leq 1\}$ and $\|f\|_2$ denotes the L_2 norm of f. Hence we need to estimate $\|T\|$.

It can be seen that, for g as in (3.101) the operator T is compact and positive; thus if α is the largest eigenvalue of T, then $\alpha = \|T\|$. Hence (3.104) (and thereby (3.103)) converges for all $\lambda \leq \alpha^{-1}$. The standard numerical methods of calculating eigenvalues show that $\alpha > 5.718$ and thus, if $\lambda \leq 1/5.718 = 0.174$, then (3.103) converges.

Comparing the branching process as in Theorem 3.2 with the Boolean model, it is obvious that the expected number of balls in a component is at most the expected total number of members in this branching process. Thus if $\lambda \leq 0.174$, then the expected number of balls in a component is finite; i.e. $\lambda_c \geq 0.174$.

To obtain the upper bound, we compare the Boolean model with a site-percolation model on the triangular lattice. Consider the triangular lattice as in Figure 3.2 with each edge being of unit length. Each site of the lattice is enclosed in a 'flower' which is formed by the six arcs of circles, each of unit radius and centred at the midpoints of the six edges adjacent to the site. A site will be called occupied if there is a point of the Poisson process X situated inside the interior of the associated flower of the site. It is clear that if there are two adjacent sites both occupied, then each of the flowers of these two sites must contain the centre of at least one ball, and because of the size of the flower, the balls centred at these flowers of radius 1 must have pairwise non-empty intersection. Thus if site percolation of the triangular lattice occurs (i.e., if there is an infinite chain of adjacent occupied sites), then percolation occurs in the Boolean model.

Now, by the construction of the site-percolation model, since the occupancy of a site depends on the realisation of the Poisson process X inside the interior of its associated flower, and for two different sites, the interiors of their respective flowers are disjoint, the occupancy of a site is independent of the occupancy of other sites and the probability that a site is occupied is $p := 1 - \exp(-\lambda A)$, where A denotes the area of a flower.

We know from the theory of discrete percolation (see Kesten 1982, p. 52) that if $p > \frac{1}{2}$ then with positive probability there is percolation in the site percolation model on the triangular lattice. Thus if $A\lambda > \log 2$, then there is percolation with positive probability in the Boolean model. An elementary calculation shows that $A \approx 0.8227$, thus we have that $\lambda_c < 0.843$.   □

It is clear that computations like this become completely untractable in higher dimensions. However, it is often easier to obtain asymptotic results for important quantities in high dimensions than to perform explicit computations in, say, two dimensions. The reason for this is that most estimates using branching processes become more and more precise when the dimension gets higher, because there is less dependency between the branches of the process in high dimensions. We shall illustrate this idea with asymptotic results on the cluster-size distribution. See the notes for more information on the asymptotics of the critical density. The following lemma is what makes high dimensions 'special':

Lemma 3.10 *Suppose that $X(d)$ and $Y(d)$ are independent and uniformly distributed on the unit ball in $\mathbb{R}^d$. Then we have*

$$\lim_{d \to \infty} P\left\{|X(d)| > \tfrac{3}{4}\right\} = 1, \tag{3.105}$$

$$\lim_{d \to \infty} \left\{ \sup_{|x| > 3/4} \{P\{|X(d) - x| \le 1\}\} \right\} = 0, \tag{3.106}$$

and

$$\lim_{d \to \infty} P\{|X(d) - Y(d)| \le 1\} = 0. \tag{3.107}$$

Proof Equation (3.105) is obvious. For (3.106), we write $|X(d)-x|^2 = |x|^2 + |X(d)|^2 - 2|X(d) \cdot x|$. Using (3.105), it is not hard to see that it suffices to prove that $X(d) \cdot x$ converges to zero in probability, uniformly on $\{x : \frac{3}{4} \le x \le 2\}$. Writing $X_1(d)$ for the first coordinate of $X(d)$, we see that it suffices to show that $|X_1(d)|$ converges to zero in probability. This follows from the fact that $E(|X_1(d)|^2) \le 1/d$. Equation (3.107) follows by conditioning on $Y(d)$ and using (3.105) and (3.106). $\qquad\square$

It will be convenient to assume that we always have a point at the origin. Also, we reparametrise the model writing the density as $\lambda(\pi_d)^{-1}$ rather than λ, where π_d denotes the volume of a ball in d dimensions with radius 1. In high dimensions it is convenient to consider the Boolean model with fixed radius $\frac{1}{2}$ rather than with radius 1. The reason for this will become clear in the proof of the following theorem.

Theorem 3.11 *Consider a Boolean model $(X, \frac{1}{2}, \lambda(\pi_d)^{-1})$ in d dimensions. Denote by $f_k(\lambda)$ the probability that a Galton–Watson branching process with Poisson-λ offspring distribution has total progeny k. Then, writing W_d for the component containing the origin in d dimensions, we have*

$$\lim_{d \to \infty} P_{\lambda(\pi_d)^{-1}}(X(W_d) = k) = f_k(\lambda).$$

Proof The important tool to use here is a stochastic process which is called a *branching random walk* (BRW). This is a random process $\{Z_n^d ; n = 0, 1, 2, \ldots\}$ such that Z_0^d consists of just the origin and Z_{n+1}^d is obtained from Z_n^d by replacing each point $y \in Z_n^d$ by an independent Poisson-λ number of points uniformly distributed in $S(y, 1)$, i.e. the unit ball centred at y. The set Z_n^d is referred to as the n-th generation of the BRW. We can order all points in this BRW as follows: all members of an earlier generation precede all members of a later generation, and the members of any particular generation are ranked in increasing distance to the origin. Let $x_1, x_2, \ldots$ be the ordering of the points of the BRW.

We modify the BRW according to the following algorithm: having checked $x_1, \ldots, x_k$, we discard x_{k+1} if (a) it is a descendant of a point previously discarded or (b) $x_{k+1} \in \cup_{j=i}^{l-1} S(x_j, 1)$, where l is such that x_l is the immediate forbear of x_{k+1}. It is not hard to see that the remaining points of the BRW have the same spatial distribution as the Poisson points forming the occupied component of the origin in the Boolean model with density $\lambda(\pi_d)^{-1}$. This coupling shows that

$$\sum_{j=1}^{k} P_{\lambda(\pi_d)^{-1}}(X(W_d) = j) - \sum_{j=1}^{k} f_j(\lambda) \ge 0, \qquad (3.108)$$

for all $k \geq 1$. The left-hand side in (3.108) is the probability that in the construction above, (i) the BRW has total progeny greater than k, and (ii) after modifying the BRW as indicated, we are left with at most k points. Denote by E_k^d the event that none of the first k points in the ordering of the BRW is thrown away. It suffices to prove that $\lim_{d \to \infty} P(E_k^d) = 1$ for all k. To do this, let x_i be the i-th point in the BRW and write F_i^d for the event that (i) none of the offspring of x_i lies in $\cup_{j=1}^{i-1} S(x_j, 1)$, (ii) no two points of this offspring are separated by a distance less than 1 and (iii) each point of this offspring is outside $S(x_i, 3/4)$. It is not hard to see that $\cap_{i=1}^{k-1} F_i^d \subset E_k^d$. From Lemma 3.10 it follows easily that both $P(F_1^d)$ and $P(F_k^d \mid F_{k-1}^d)$ tend to 1 when d tends to infinity, and the proof is complete. □

3.10 Notes

The question of complete coverage of a region $A \subseteq \mathbb{R}^d$ by a Boolean model is one of the oldest problems in stochastic geometry, more details of which may be obtained in Hall (1988).

The material of Sections 3.2 and 3.3 and the bounds for λ_c in two dimensions in Section 3.9 are contained in Hall (1985), which is one of the first mathematical papers devoted exclusively to the study of continuum percolation. The results in Sections 3.4 and 3.7 are from Roy (1990), which also contains the equality result of Section 3.5 but for two dimensions only. The d-dimensional equality result is a combination of results of Zuev and Sidorenko (1985), Menshikov (1986) and Meester, Roy and Sarkar (1994). It may be pointed out here that the results of Zuev and Sidorenko (1985) and Menshikov (1986) establish the equality only in the case when the radius random variable is bounded from below by a strictly positive quantity. The result of Meester, Roy and Sarkar (1994) is used here to extend this case to an arbitrary bounded positive radius random variable.

The uniqueness result of Section 3.6 is from Meester and Roy (1994). Continuity and convergence results of Section 3.8 are from Meester, Roy and Sarkar (1994) and Sarkar (1994). It may be noted that Penrose (1995c) generalised some of these results to a wider class of radii distributions, using results of Tanemura (1993). The latter paper contains a continuum version of a renormalisation technique of Grimmett and Marstrand (1990) for discrete nearest-neighbour percolation models. The asymptotics in Section 3.9 are taken from Penrose (1995b). The method using the branching random walk can be pushed much further. In the same paper, Penrose shows that the critical density $\lambda_c^{(d)}$ in dimension d, satisfies $\lim_{d \to \infty} \pi_d \lambda_c^{(d)} = 1$. This means that if the expected number of balls intersecting the unit ball at the origin is larger than 1, the model percolates with positive probability in high dimensions. This corresponds with positive survival probability of a Galton–Watson branching process when the expected number of offspring is larger than 1.

4

Vacancy in Poisson Boolean models

In this chapter we discuss the properties of the vacant components in the Poisson Boolean model. Unlike the occupied components in the Poisson Boolean model, where the structure of the occupied regions arises because of placing balls around the Poisson points, the vacancy structure arises in the negative sense, i.e., in the absence of any ball covering a point. This lack of a structure to describe directly the vacancy configuration is a limitation due to which it is often harder to establish results concerning vacancy.

In the study of percolation on discrete graphs, the vacancy configuration is usually thought of as the 'dual' of the occupancy structure. In that sense we shall occasionally refer to the vacant region as the dual of the occupied region. This nomenclature is more informal than exact, because in the discrete percolation models, the dual structure has a legitimate construct of its own, rather than being just an appendage of the occupied structure.

We shall define critical densities via vacancy and show that in two dimensions, when the radii are bounded, $\lambda_H^* = \lambda_T^*$, where these notations have the same meaning in the vacancy as they had (without the superscript) in the occupancy. In addition, we shall show that in two dimensions, the critical densities arising from the occupancy agree with that arising from vacancy. We shall also establish a uniqueness result for the vacant component as in Section 3.6 of Chapter 3. This will show that in two dimensions for $\lambda < \lambda_c$, with probability 1, there is exactly one unbounded vacant component and no unbounded occupied component, and for $\lambda > \lambda_c$, with probability 1, there is exactly one unbounded occupied component and no unbounded vacant component. For three and higher dimensions, we do not have a result establishing the equality of λ_H^* and λ_T^*, although it is expected to be true. We also do not have any result to show the co-existence of unbounded vacant and occupied components – i.e.,

$\lambda_H < \lambda_H^*$ – in three or higher dimensions as suggested by simulation studies in physics.

To prove the equality in two dimensions, we establish an RSW lemma for the Poisson Boolean model. Russo (1978) and Seymour and Welsh (1978) independently proved that in a $\mathbf{Z}^2$ lattice-percolation model, if the crossing probabilities of suitable rectangles in either direction are larger than δ, then the crossing probability of a bigger rectangle is larger than $f(\delta)$, where the function f depends only on the ratio of the size of the larger and the smaller rectangle. The idea of their proof is to connect a left–right crossing of a rectangle $[0, l_1] \times [0, l_2]$ and a left–right crossing of another rectangle $[l_1/2, 3l_1/2] \times [0, l_2]$ by a top–bottom crossing of a suitable rectangle to yield a left–right crossing of the rectangle $[0, 3l_1/2] \times [0, l_2]$. Using similar ideas and employing a lattice approximation of the continuum model, we prove the RSW lemma for vacant crossings in our Poisson Boolean model. However, we need some additional construction to take care of the dependency structure of our model. As we shall see later, this involves modifying the Poisson Boolean model in a small region near the intersection of the lowest vacant left–right crossing of $[0, l_1] \times [0, l_2]$ with the right edge of the rectangle $[0, l_1] \times [0, l_2]$ to obtain a dependence structure where we can apply the FKG inequality.

4.1 Critical densities

As in the case of the occupied region, the size of a vacant cluster can be measured by either its diameter or its Lebesgue measure. The notion of measuring the size of an occupied cluster by the number of Poisson points comprising the cluster does not have any analogue in vacancy. Thus we have the critical densities:

$$\lambda_c^* := \sup\{\lambda : P_\lambda\{d(V) = \infty\} > 0\},$$
$$\lambda_D^* := \sup\{\lambda : E_\lambda(d(V)) = \infty\},$$
$$\lambda_H^* := \sup\{\lambda : P_\lambda\{\ell(V) = \infty\} > 0\},$$
$$\lambda_T^* := \sup\{\lambda : E_\lambda(\ell(V)) = \infty\},$$

where V is the vacant component of the origin as defined in Section 1.4.

In addition to the above four critical densities, we have the critical density defined through vacant crossings:

$$\lambda_S^* := \sup\{\lambda : \limsup_{n \to \infty} \sigma^*((n, 3n, \ldots, 3n), \lambda, 1) > 0\},$$

where $\sigma^*((n, 3n, \ldots, 3n), \lambda, 1)$ is the probability, under P_λ, of the existence of a vacant crossing in the short direction of the rectangle $[0, n] \times [0, 3n] \times \cdots \times [0, 3n]$ as defined in Section 2.3.

Our first goal is to show that, when ρ is bounded, the critical densities corresponding to the two different notions of measuring the size of a vacant cluster are equal.

Theorem 4.1 *For a Poisson Boolean model (X, ρ, λ) on $\mathbb{R}^d$ with ρ bounded almost surely, we have (a) $\lambda_c^* = \lambda_H^*$ and (b) $\lambda_D^* = \lambda_T^*$.*

Unlike the proof of the analogous result for the occupancy (Theorem 3.4), for the proof of Theorem 4.1 we need a bound on the growth of the vacant cluster V when the probability of a vacant crossing is very small. The following lemma can be proved as Lemma 3.3.

Lemma 4.1 *Let (X, ρ, λ) be a Poisson Boolean model on $\mathbb{R}^d$ with $0 < \rho \leq R$ a.s. for some $R > 0$. There exists $\kappa_0 > 0$ such that, if for some $N = (N_1, \ldots, N_d)$ with $N_j \geq R$, for all $1 \leq j \leq d$, we have*

$$\sigma^*((3N_1, \ldots, 3N_{i-1}, N_i, 3N_{i+1}, \ldots, 3N_d), \lambda, i) \leq \kappa_0, \qquad (4.1)$$

for all $1 \leq i \leq d$, then

$$P_\lambda\{d(V) \geq a\} \leq C_1 \exp(-C_2 a) \qquad (4.2)$$

and

$$P_\lambda\{\ell(V) \geq a\} \leq C_3 \exp(-C_4 a), \qquad (4.3)$$

for all $a > 0$, where C_1, C_2, C_3 and C_4 are positive constants independent of a.

An immediate consequence of this lemma is that for a Poisson Boolean model with a bounded radius random variable, we have

$$\lambda_D^* \leq \lambda_S^* \quad \text{and} \quad \lambda_T^* \leq \lambda_S^*. \qquad (4.4)$$

Proof of Theorem 4.1 For ease of notation we shall present the proof only for two dimensions. As can be seen from the proof, the extension to higher dimensions is straightforward.

First we show (b) $\lambda_D^* = \lambda_T^*$. Let $B_{2R}(i) = (0, i4R) + B_{2R}$ for all $i \geq 0$. Observe that an L–R vacant crossing of the rectangle $[0, 3^k] \times [0, 3^{k+1}]$ necessitates the existence of a vacant region starting from the left edge of $[0, 3^k] \times [0, 3^{k+1}]$

which has a diameter at least 3^k. Hence we have, for any $k \geq 1$,

$$\sigma^*((3^k, 3^{k+1}), \lambda, 1)$$

$$\leq P_\lambda \left(\bigcup_{i=0}^{3^{k+1}/4R} \{d(V(B_{2R}(i))) \geq 3^k\} \right)$$

$$\leq \sum_{i=0}^{3^{k+1}/4R} P_\lambda\{d(V(B_{2R}(i))) \geq 3^k\}$$

$$\leq \left(\frac{3^{k+1}}{4R} + 1 \right) P_\lambda\{d(V(B_{2R}(0))) \geq 3^k\}. \tag{4.5}$$

In the preceding inequalities we have assumed, without any loss of generality, that $3^{k+1}/4R$ is an integer.

If $\lambda > \lambda_D^*$, then $E_\lambda(d(V)) < \infty$ and thus

$$\sum_{k=1}^{\infty} 3^{k+1} P_\lambda(d(V) \geq 3^k) \leq C_1 E_\lambda(d(V)) < \infty, \tag{4.6}$$

where C_1 is a positive constant. Now as in Example 2.1 of Section 2.3, from (4.6) we have

$$\sum_{k=1}^{\infty} 3^{k+1} P_\lambda(d(V(B_{2R}(0))) \geq 3^k) \leq C_2 E_\lambda(d(V(B_{2R}(0)))) < \infty, \tag{4.7}$$

for some constant $C_2 > 0$. Thus from (4.7) and (4.5) we have

$$\sum_{k=1}^{\infty} \sigma^*((3^k, 3^{k+1}), \lambda, 1) < \infty. \tag{4.8}$$

Hence, for some integer $k_0 \geq 0$ we have $\sigma^*((3^k, 3^{k+1}), \lambda, 1) \leq \kappa_0$ for all $k \geq k_0$, where κ_0 is as in Lemma 4.1. Thus applying that lemma, we obtain $\lambda \geq \lambda_T^*$. This shows that $\lambda_T^* \leq \lambda_D^*$.

To complete the proof of (b) we need to show that $\lambda_D^* \leq \lambda_T^*$. We use an argument based on scaling. We first show this in the case when there exists $\eta > 0$ such that $\rho \geq \eta$ a.s. Fix $\lambda < \lambda_D^*$ and consider the Poisson Boolean models $(X_a, a\rho, \lambda)$, for $0 < a \leq 1$. We couple these models on the same probability space and thus, letting V_a denote the vacant cluster of the origin in $(X_a, a\rho, \lambda)$, we have $V_b \subseteq V_a$ for all $0 < a \leq b \leq 1$. Moreover, if $x \in V_1$, then, for any $0 < a < 1$, the open ball centred at x of radius $(1 - a)\eta$ will be completely contained in V_a, and so

$$\ell(V_a) \geq \pi((1-a)\eta)^2 \frac{d(V_1)}{2(1-a)\eta}$$

$$= C(a)d(V_1),$$

for some constant $C(a) > 0$. Hence, if $d(V) = \infty$ then $\ell(V_a) = \infty$. This implies that whenever $\lambda < \lambda_D^*(\rho)$ we have

$$\lambda \leq \lambda_T^*(a\rho), \quad \text{for all } a > 0. \tag{4.9}$$

However, from Proposition 2.11 we have $\lambda_T^*(a\rho) = \lambda_T^*(\rho)/a^d$, so (4.9) yields the desired inequality.

For general ρ we use an argument similar to the one used in Case 2 of the proof of (a) in Theorem 3.4, and as such we omit it.

Next we show (a) $\lambda_c^* = \lambda_H^*$. We note that for $\lambda < \lambda_H^*$ and for any $m > 0$,

$$\begin{aligned}
P_\lambda\{d(V) = \infty\} &\geq \lim_{m \to \infty} P_\lambda\{V \cap B_m^c \neq \emptyset\} \\
&\geq \lim_{m \to \infty} P_\lambda\{\ell(V) \geq (2m)^2\} \\
&\geq P_\lambda\{\ell(V) = \infty\}.
\end{aligned}$$

Thus, if $\lambda < \lambda_H^*$ then $\lambda \leq \lambda_c^*$.

Now suppose $\lambda < \lambda_c^*$; i.e., $P_\lambda\{d(V) = \infty\} > 0$. We partition the space with boxes of the form $B_m^{2mz} = 2mz + B_m$, where $z \in \mathbf{Z}^d$. Consider the event $F_z = \{B_m^{2mz}$ is completely contained in an unbounded vacant component$\}$. It is clear that $P(F_z)$ is positive and independent of z. The translation over the vector $2me_1$ is ergodic, and it follows by the ergodic theorem that $P(F_z \text{ i.o.}) = 1$. But then, using the uniqueness of the unbounded component (see Section 4.8) we conclude that for all z for which F_z occurs, the box B_m^{2mz} is contained in the same unbounded component V'. It follows that $\ell(V') = \infty$, whence $\lambda < \lambda_H^*$. This establishes that $\lambda_c^* = \lambda_H^*$. □

REMARK: It is possible to prove the last step of the preceding proof; i.e. $\lambda_c^* \leq \lambda_H^*$, along the same lines as the proof of (b); this proof does not involve uniqueness. However, the preceding proof is presented for its elegance and because it does not use the boundedness of the radii.

4.2 RSW – notation and definition

Consider a Poisson Boolean model (X, ρ, λ) on $\mathbb{R}^2$. We define a continuous curve γ to be a *vacant path* if $\gamma \subset C^c$, the vacant region. A continuous curve $\tilde{\gamma}$ is said to be an *occupied path* if $\tilde{\gamma} \subset C$. A vacant path γ is said to be a *vacant left–right (L–R) crossing* of the rectangle $[0, l_1] \times [0, l_2]$ if $\gamma \cap (\{0\} \times [0, l_2]) \neq \emptyset$, $\gamma \cap (\{l_1\} \times [0, l_2]) \neq \emptyset$ and, except for its end points, γ is contained in $(0, l_1) \times (0, l_2)$. Similarly, we define a *vacant top–bottom (T–B) crossing* by requiring that the end points of γ lie on the top and bottom edges of the rectangle, respectively. We assume that all curves under consideration do not have any self-intersections.

Let $\mathbb{L}_n$ be the lattice $a_n \mathbf{Z} \times a_n \mathbf{Z}$, where $\{a_n\}$ is a sequence of positive numbers decreasing to zero when n tends to infinity. Suppose l_1 and l_2 are positive integer multiples of a_n. By a cell in this lattice we mean a set $[a_n z_1, a_n z_1 + a_n] \times [a_n z_2, a_n z_2 + a_n]$ for $z_1, z_2 \in \mathbf{Z}$. Note here that we include the perimeter of the cell in the definition. Two cells in the lattice are said to be *adjacent* if they have an edge in common. A cell C_0 in $\mathbb{L}_n$ is called *vacant* (respectively, *occupied*) if $C_0 \cap (\cup_{i \geq 1} S(x_i)) = \emptyset$ (respectively, $C_0 \cap (\cup_{i \geq 1} S(x_i)) \neq \emptyset$). An $\mathbb{L}_n$-*path* is a sequence of disjoint adjacent cells. A *vacant (occupied) $\mathbb{L}_n$-path* is an $\mathbb{L}_n$-path which consists of only vacant (occupied) adjacent cells. An L–R $\mathbb{L}_n$-crossing Γ of the rectangle $[0, l_1] \times [0, l_2]$ is an $\mathbb{L}_n$-path such that $\Gamma \subseteq [0, l_1] \times [0, l_2]$, $\Gamma \cap (\{0\} \times [0, l_2]) \neq \emptyset$, $\Gamma \cap (\{l_1\} \times [0, l_2]) \neq \emptyset$ and, in addition, each of $\Gamma \cap (\{0\} \times [0, l_2])$ and $\Gamma \cap (\{l_1\} \times [0, l_2])$ consists of a single edge. We define T–B $\mathbb{L}_n$-crossings, vacant/occupied L–R $\mathbb{L}_n$-crossings and vacant/occupied T–B $\mathbb{L}_n$-crossings in a similar fashion. The crossing probabilities (see (2.19)) are denoted as follows:

$$\sigma^*((l_1, l_2), \lambda, 1)$$
$$= P_\lambda\{\text{there exists a vacant L–R crossing of } [0, l_1] \times [0, l_2]\},$$

$$\sigma^*((l_1, l_2), \lambda, 2)$$
$$= P_\lambda\{\text{there exists a vacant T–B crossing of } [0, l_1] \times [0, l_2]\}.$$

The RSW lemma states the following:

Theorem 4.2 (RSW lemma) *Let (X, ρ, λ) be a Poisson Boolean model in two dimensions with*

$$0 < \rho \leq R \text{ a.s.} \tag{4.10}$$

for some $R > 0$. If there exist constants $\delta_1 > 0$ and $\delta_2 > 0$ such that

$$\sigma^*((l_1, l_2), \lambda, 1) \geq \delta_1 \tag{4.11}$$

and

$$\sigma^*((l_3, l_2), \lambda, 2) \geq \delta_2, \tag{4.12}$$

for some $l_1 > 4R$ and $2R < l_3 \leq 3l_1/2$, then for any integer k,

$$\sigma^*((kl_1, l_2), \lambda, 1) \geq C_k(\lambda, R) f_k(\delta_1, \delta_2), \tag{4.13}$$

where $C_k(\lambda, R) > 0$ is independent of δ_1 and δ_2 and $f_k(\delta_1, \delta_2) > 0$.

The proof of this theorem is presented in the next three sections and applications can be found in Section 4.6. The main step of the proof is to obtain an RSW result for a discrete approximation of the Boolean model. This, however, does not follow from the discrete percolation results because in any discretisation of the Boolean model, the dependency structure of the model remains. We have to warn the reader that the proof of the RSW lemma is very technical; one can safely skip the next three sections without disturbing the flow of the book.

4.3 RSW – construction

We take n so large that

$$l_1 > 4R + a_n. \tag{4.14}$$

Let $l_{1n} := \lfloor l_1/4a_n \rfloor 4a_n$ and $l_{3n} := \lfloor l_3/a_n \rfloor a_n + a_n$, where for any $x \in \mathbb{R}$, $\lfloor x \rfloor$ denotes the largest integer less than or equal to x. Clearly, both l_{1n} and l_{3n} are integer multiples of a_n. By the monotonicity property of crossing probabilities, $\sigma^*((l_{1n}, l_2), \lambda, 1) \geq \sigma^*((l_1, l_2), \lambda, 1)$ and $\sigma^*((l_{3n}, l_2), \lambda, 2) \geq \sigma^*((l_3, l_2), \lambda, 2)$. Moreover, as $a_n \to 0$, $l_{1n} \to l_1$ and $l_{3n} \to l_3$ and it is easy to check that also $\sigma^*((l_{1n}, l_2), \lambda, 1) \to \sigma^*((l_1, l_2), \lambda, 1)$ and $\sigma^*((l_{3n}, l_2), \lambda, 2) \to \sigma^*((l_3, l_2), \lambda, 2)$. Since in our calculations we later let $a_n \to 0$, we may replace l_{1n} by l_1 and l_{3n} by l_3 and thus for the sake of simplicity in notation, we assume that in addition to (4.14) the following also holds:

$$R/4, l_1/4, l_2 \text{ and } l_3 \text{ are all integer multiples of } a_n. \tag{4.15}$$

On the lattice $\mathbb{L}_n$ of 'size a_n' we fix an L–R self-avoiding (i.e. all cells in the path are different) $\mathbb{L}_n$-crossing r which consists of cells $C_0, C_1, \ldots, C_m$ of $[0, l_1 - R] \times [0, l_2]$ with $C_0 \cap (\{0\} \times [0, l_2]) \neq \emptyset$, $C_m \cap (\{l_1 - R\} \times [0, l_2]) \neq \emptyset$, $\{C_1, \ldots, C_{m-1}\} \subseteq (0, l_1 - R) \times [0, l_2]$ and C_i and C_{i+1} are adjacent for all $i = 0, \ldots, m - 1$. We now consider different pieces of the $\mathbb{L}_n$-crossing r. Suppose $0 \leq i_1 < i_2 < \cdots < i_{l(r)} \leq m$ are all indices such that $C_i \cap (\{l_1/4\} \times [0, l_2]) \neq \emptyset$ and $C_i \subseteq [l_1/4, l_1 - R] \times [0, l_2]$. In other words C_{i_j}, $j = 1, 2, \ldots, l(r)$ are all the cells of r which are adjacent to the line $\{l_1/4\} \times [0, l_2]$ and lie on the right side of this line. By our choice of notation, $C_{i_{l(r)}}$ is the 'last intersection' of r with $\{l_1/4\} \times [0, l_2]$, i.e., $C_j \subset (l_1/4, l_1 - R] \times [0, l_2]$ for all $j > i_{l(r)}$. Let $F(r) := (C_{i_{l(r)}}, C_{i_{l(r)}+1}, \ldots, C_m)$ be the piece of r after this last intersection. Let $r_k := (C_{i_k}, C_{i_k+1}, \ldots, C_{i_{k+1}})$ for all $k = 1, \ldots, l(r) - 1$. Also let $1 \leq j_1 \leq j_2 \leq \cdots \leq j_{b(r)} \leq l(r) - 1$ be such that $r_j \subseteq [l_1/4, l_1 - R] \times [0, l_2]$ if and only if $j \in \{j_1, j_2, \ldots, j_{b(r)}\}$. Thus $F(r)$ is the piece of r lying completely in the rectangle $[l_1/4, l_1 - R] \times [0, l_2]$ and on the right of the line $\{l_1/4\} \times [0, l_2]$, with one of its end cells adjacent to

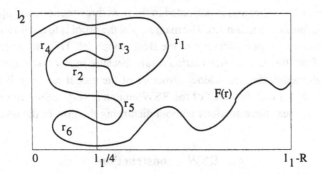

Figure 4.1. The paths r_1, r_3 and r_5 are the paths r_{j_1}, r_{j_2} and r_{j_3}.

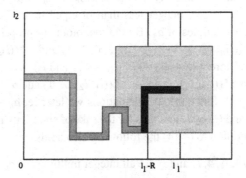

Figure 4.2. The shaded path is r, the blackened, Γ_r and the dotted region, A'_r.

the line $\{l_1/4\} \times [0, l_2]$. Also, for every $1 \leq i \leq b(r) - 1$, r_{j_i} has both its end cells adjacent to the line $\{l_1/4\} \times [0, l_2]$ and lies completely in the rectangle $[l_1/4, l_1 - R] \times [0, l_2]$. (Figure 4.1 depicts this notation.) Let the corner vertices of the cell C_0 be $(0, \alpha)$, (a_n, α) $(a_n, \alpha + a_n)$ and $(0, \alpha + a_n)$ and the corner vertices of the cell C_m be $(l_1 - R, \beta)$, $(l_1 - R - a_n, \beta)$, $(l_1 - R - a_n, \beta + a_n)$ and $(l_1 - R, \beta + a_n)$. Let Γ_r be an $\mathbb{L}_n$-path defined as follows (see Figure 4.2): Γ_r is the collection of cells in the region

(i) $([l_1 - R - a_n, l_1 - R] \times [\beta, \beta + R + 2a_n]) \cup ([l_1 - R, l_1] \times [\beta + R + a_n, \beta + R + 2a_n])$ if $\beta < l_2 - R - 2a_n$,

(ii) $([l_1 - R - a_n, l_1 - R] \times [\beta, l_2]) \cup ([l_1 - R, l_1] \times [l_2 - a_n, l_2])$ if $\beta \geq l_2 - R - 2a_n$.

The $\mathbb{L}_n$-path $r \cup \Gamma_r$ is an $\mathbb{L}_n$-crossing of $[0, l_1] \times [0, l_2]$ (see Figure 4.2).

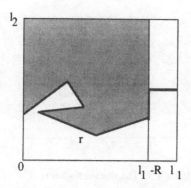

Figure 4.3. The shaded region is J_1^+.

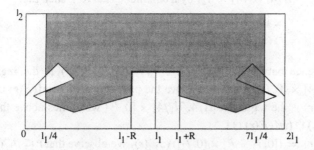

Figure 4.4. The shaded region is J_2^+.

For any set of cells s in $[0, l_1] \times [0, l_2]$, let ref(s) be the reflection of s in $\{l_1\} \times [0, l_2]$. Then $(F(r) \cup \Gamma_r) \cup (\text{ref}(F(r) \cup \Gamma_r))$ is an L–R $\mathbb{L}_n$-crossing of $[l_1/4, 7l_1/4] \times [0, l_2]$. Next we define the region (see Figure 4.3)

$$J_1^+(r) := \{(x, y) \in [0, l_1 - R] \times [0, l_2] : (x, y) \text{ can be connected to}$$
$$[0, l_1 - R] \times \{l_2\} \text{ by a continuous curve } \gamma \text{ such that}$$
$$\gamma \subseteq [0, l_1 - R] \times [0, l_2] \text{ and } \gamma \cap r = \emptyset\}.$$

This is the part of $[0, l_1 - R] \times [0, l_2]$ which lies above r. We also define the regions (see Figures 4.4 and 4.5)

$$J_2^+(r) := \{(x, y) \in [l_1/4, 7l_1/4] \times [0, l_2] : (x, y) \text{ can be connected to}$$
$$[l_1/4, 7l_1/4] \times \{l_2\} \text{ by a continuous curve } \gamma \text{ such that}$$
$$\gamma \subseteq [l_1/4, 7l_1/4] \times [0, l_2]$$
$$\text{and } \gamma \cap ((r \cup \Gamma_r) \cup (\text{ref}(F(r) \cup \Gamma_r))) = \emptyset\}.$$

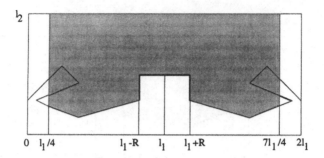

Figure 4.5. The shaded region is J_3^+.

$J_3^+(r) := \{(x, y) \in [l_1/4, 7l_1/4] \times [0, l_2] : (x, y)$ can be connected to
$[l_1/4, 7l_1/4] \times \{l_2\}$ by a continuous curve γ such that
$\gamma \subseteq [l_1/4, 7l_1/4] \times [0, l_2]$ and
$\gamma \cap ((F(r) \cup \Gamma_r) \cup (\text{ref}(F(r) \cup \Gamma_r))) = \emptyset\}$.

The difference between $J_2^+(r)$ and $J_3^+(r)$ is that $J_2^+(r)$ is the region in
$[l_1/4, 7l_1/4] \times [0, l_2]$ which lies above the path $(r \cup \Gamma_r) \cup (\text{ref}(F(r) \cup \Gamma_r))$,
while $J_3^+(r)$ is the region in $[l_1/4, 7l_1/4] \times [0, l_2]$ which lies above the path
$(F(r) \cup \Gamma_r) \cup (\text{ref}(F(r) \cup \Gamma_r))$.

Let $J_1^-(r) := ([0, l_1 - R] \times [0, l_2]) \backslash J_1^+(r)$. We observe that $r \subseteq J_1^-(r)$ and
$J_1^+(r)$ is a connected region, while $J_2^+(r)$ is connected if $\beta < l_2 - R - 2a_n$.
For any region A, let $\overline{A}$ denote the closure of A, and $\text{int}(A)$ the interior.

For r as above let A_r' be the region $([l_1 - 2R - a_n, l_1 + R] \times [\beta, \beta + 2R + 2a_n]) \cap \overline{(J_1^+(r)} \cup ([l_1 - R, \infty) \times \mathbb{R}) \cup (\mathbb{R} \times [l_2, \infty)))$ (see Figure 4.2). We introduce the following events:

$A_r := \{X(A_r') = 0\}$,

$E_r := \{r$ is vacant$\}$,

$L_r := \{$any L–R $\mathbb{L}_n$-crossing s of $[0, l_1 - R] \times [0, l_2]$ such that
$\quad s \neq r$ and $s \subseteq J_1^-(r)$ is not vacant,

$D_2(r) := \{$there exists a vacant $\mathbb{L}_n$-path $s'' := (C_0'', C_1'', \ldots, C_\nu'')$ such that
$\quad s'' \subseteq \overline{J_2^+(r)}, \; C_0'' \cap ([l_1/4, 7l_1/4] \times \{l_2\}) \neq \emptyset$,
$\quad C_\nu'' \cap (F(r) \cup \Gamma_r) \neq \emptyset$ and
$\quad C_k'' \cap (([l_1/4, 7l_1/4] \times \{l_2\}) \cup (F(r) \cup \Gamma_r)) = \emptyset$,
$\quad$ for all $k = 1, \ldots, \nu - 1\}$,

$D_3(r) := \{$there exists a vacant $\mathbb{L}_n$-path $s' := (C_0', C_1', \ldots, C_\mu')$

such that $s' \subseteq \overline{J_3^+(r)}$, $C_0' \cap ([l_1/4, 7l_1/4] \times \{l_2\}) \neq \emptyset$,

$C_\mu' \cap (F(r) \cup \Gamma_r) \neq \emptyset$ and

$C_k' \cap (([l_1/4, 7l_1/4] \times \{l_2\}) \cup (F(r) \cup \Gamma_r)) = \emptyset$

for all $k = 1, \ldots, \mu - 1\}$,

$D(r) := \{$there exist disjoint vacant $\mathbb{L}_n$-paths $s_1, \ldots, s_q$ for some

$q \in \{1, \ldots, b(r)\}$ with $s_k := (C_{0,k}, \ldots, C_{\mu_k,k})$ for all

$k = 1, \ldots, q$, such that

(i) $C_{0,1} \cap ([l_1/4, 7l_1/4] \times \{l_2\}) \neq \emptyset$, $C_{\mu_q,q} \cap (F(r) \cup \Gamma_r) \neq \emptyset$,

(ii) $C_{i,k} \subseteq \overline{J_2^+(r)}$ for all $i = 1, \ldots, \mu_k - 1$ and for all

$k = 1, \ldots, q$,

(iii) for all $k = 1, \ldots, q - 1$, $C_{\mu_k,k}$ lies on r_{i_k} and $C_{0,k+1}$ lies

on $r_{i_{k'}}$, where $i_k, i_{k'} \in \{j_1, \ldots, j_{b(r)}\}$ and $i_{k'}$ is such that

either $i_{k'} = i_k$ or there exists a set of indices Φ_k in

$\{j_1, \ldots, j_{b(r)}\}$ such that $r_{i_{k'}}$ can be connected to r_{i_k}

by a continuous path γ which lies completely in

$\text{int}\,[(\cup_{i \in \Phi_k} r_i) \cap ([l_1/4, 7l_1/4] \times [0, l_2])]\}$.

In words, $D_3(r)$ is the event that there is a vacant $\mathbb{L}_n$-path in $J_3^+(r)$ which connects the top edge of the rectangle $[l_1/4, 7l_1/4] \times [0, l_2]$ to the path $F(r) \cup \Gamma_r$; $D_2(r)$ is the event that there is a vacant $\mathbb{L}_n$-path in $J_2^+(r)$ which connects the top edge of the rectangle $[l_1/4, 7l_1/4] \times [0, l_2]$ to the path $F(r) \cup \Gamma_r$ and $D(r)$ is the event that there are vacant $\mathbb{L}_n$-paths in $[l_1/4, 7l_1/4] \times [0, l_2]$ connecting some of the r_i's adjacent to the boundary of $J_2^+(r)$, and two other vacant $\mathbb{L}_n$-paths, one connecting one such r_i to the path $F(r) \cup \Gamma_r$ and the other connecting another such r_i to the top edge of the rectangle $[l_1/4, 7l_1/4] \times [0, l_2]$ (see Figures 4.6 and 4.7).

Given two L–R $\mathbb{L}_n$-crossings r_1 and r_2 of $[0, a_1] \times [0, a_2]$ (for some positive numbers a_1 and a_2) we define $r_1 \preceq r_2$ if $J_1^+(r_2) \subseteq J_1^+(r_1)$. For any two L–R $\mathbb{L}_n$-crossings r_1 and r_2, there is an L–R $\mathbb{L}_n$-crossing s in $r_1 \cup r_2$ of $[0, a_1] \times [0, a_2]$ such that $J_1^+(s) = J_1^+(r_1) \cup J_1^+(r_2)$ and so $s \preceq r_1$ and $s \preceq r_2$. In particular if r_1 and r_2 are such that $r_1 \preceq r_2$ then $s = r_1$ satisfies $s \preceq r_1$ and $s \preceq r_2$. However, if neither $r_1 \preceq r_2$ nor $r_2 \preceq r_1$ hold, then $r_1 \cap r_2 \neq \emptyset$. If we consider the L–R $\mathbb{L}_n$-crossing r' of $[0, a_1] \times [0, a_2]$ which is the 'lower' part of $r_1 \cup r_2$ then it is not hard to believe that for $s = r'$ the following is satisfied: *there exists a L–R $\mathbb{L}_n$-crossing s of $[0, a_1] \times [0, a_2]$ such that*

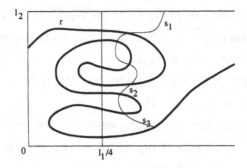

Figure 4.6. The thick line is the path **r**, the thin line is the path **s** with segments s_1, s_2 and s_3 as in the event $D(\mathbf{r})$.

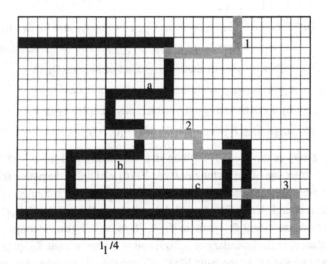

Figure 4.7. The black line is a portion of r, the segments 1, 2 and 3 are s_k, s_{k+1} and s_{k+2}, respectively, and the segments a, b and c are r_{i_k}, $r_{i_{k'}}$ and $r_{i_{k+1}}$.

(a) $s \subseteq r_1 \cup r_2$, *(b)* $s \preceq r$ *for all L–R $\mathbb{L}_n$-crossings r of* $[0, a_1] \times [0, a_2]$ *with* $r \subseteq r_1 \cup r_2$. Thus for the finite collection $C = \{$all vacant L–R $\mathbb{L}_n$-crossings of the rectangle $[0, a_1] \times [0, a_2]\}$, we can define the *lowest* vacant L–R $\mathbb{L}_n$-crossing of $[0, a_1] \times [0, a_2]$ as the L–R $\mathbb{L}_n$-crossing r of $[0, a_1] \times [0, a_2]$ such that $r \preceq s$ for every $s \in C$ and $r \subseteq \cup_{s \in C} s$. Note that all $\mathbb{L}_n$-cells comprising $\cup_{s \in C} s$ are vacant, and hence the L–R $\mathbb{L}_n$-crossing r is also a vacant L–R $\mathbb{L}_n$-crossing of $[0, a_1] \times [0, a_2]$ and thus $r \in C$. The existence of such a lowest vacant L–R

$\mathbb{L}_n$-crossing is intuitively clear; however a formal proof is quite technical and we refer the reader to Lemma 1 of Kesten (1982). Let $D_0(r) = D_2(r) \cup D(r)$. We observe:

Lemma 4.2 *The following hold:*

(i) $E_r \cap L_r = \{r$ *is the lowest vacant L–R $\mathbb{L}_n$-crossing of* $[0, l_1 - R] \times [0, l_2]\}$,

(ii) $D_3(r) \cap E_r = D_0(r) \cap E_r$.

Proof The statement of the lemma is quite easy to see from the definitions but because of the topological aspects very messy to write. We refrain from spelling out the details. □

4.4 RSW – preliminary results

In this section we shall use the FKG inequality and a conditional independence property, obtained from our choice of Γ_r, to derive some preliminary inequalities. It is to have this conditional independence that we introduced the path Γ_r instead of using an L–R crossing of $[0, l_1] \times [0, l_2]$ directly.

Let $\mathcal{R}$ denote the (random) lowest vacant L–R $\mathbb{L}_n$-crossing of $[0, l_1 - R] \times [0, l_2]$ in the Poisson Boolean model (X, ρ, λ) and r a fixed self-avoiding L–R $\mathbb{L}_n$-crossing of $[0, l_1 - R] \times [0, l_2]$.

Lemma 4.3 $P_\lambda(A_r \cap D_0(r) | \mathcal{R} = r) \geq P_\lambda(A_r) P_\lambda(D_3(r))$.

Proof First, we show that given E_r, the events L_r and $A_r \cap D_0(r)$ are conditionally independent. Indeed, any ball centred in $J_1^+(r)$ which intersects $J_1^-(r)$ must also intersect r. On E_r no such ball can exist. Moreover, for α as chosen in Section 4.3 and for any $x \in A_r' \cup ([-R, 0] \times [\alpha, \infty))$, $d(x, r) = d(x, J_1^-(r))$, where R is as in (4.10). So, any ball centred in $A_r' \cup ([-R, 0] \times [\alpha, \infty))$ which intersects $J_1^-(r)$ must also intersect r. Also, any ball of radius at most R centred in $((-\infty, -R) \times \mathbb{R}) \cup ((l_1, \infty) \times \mathbb{R}) \cup (\mathbb{R} \times (-\infty, -R))$ does not intersect $J_1^-(r)$. Thus given E_r, the event L_r depends on the Poisson points in the region $\Lambda_r := J_1^-(r) \cup ([-R, 0] \times [-R, \alpha)) \cup ([l_1 - R, l_1] \times [-R, \beta)) \cup ([0, l_1] \times [-R, 0])$, where β is as defined in Section 4.3.

We now show that, given E_r, no ball with centre in Λ_r can influence the occurrence of $D_0(r)$. Indeed, any ball of radius at most R centred in $J_1^-(r)$ which intersects $\overline{J_2^+(r)}$ must intersect r. Also, no ball of radius at most R centred in either $[-R, 0] \times [-R, \alpha]$ or $[0, l_1] \times [-R, 0]$ can intersect

$\overline{J_2^+(r)}$ without intersecting r. Now, for any $y \in [l_1 - R, l_1] \times [-R, \beta]$, $d(y, r) \leq d(y, \overline{J_2^+(r)})$, so any ball of radius at most R with centre in $[l_1 - R, l_1] \times [-R, \beta]$ which intersects $\overline{J_2^+(r)}$ must intersect r. Again, on E_r no such ball can exist and therefore no ball in Λ_r can influence the occurrence of $D_0(r)$. Thus, conditioned on E_r, the event L_r depends on the Poisson points inside the region Λ_r, while $(A_r \cap D_0(r))$ depends on the Poisson points outside Λ_r. Hence, given E_r, the events L_r and $(A_r \cap D_0(r))$ are independent. It follows that

$$P_\lambda(A_r \cap D_0(r) | \mathcal{R} = r) = P_\lambda(A_r \cap D_0(r) | E_r \cap L_r)$$

$$= \frac{P_\lambda(A_r \cap D_0(r) \cap L_r | E_r)}{P_\lambda(L_r | E_r)}$$

$$= \frac{P_\lambda(A_r \cap D_0(r) | E_r) P_\lambda(L_r | E_r)}{P_\lambda(L_r | E_r)}.$$

Now using (ii) of Lemma 4.2 and noting that all the three events A_r, E_r and $D_3(r)$ are decreasing, we obtain from the FKG inequality,

$$P_\lambda(A_r \cap D_0(r) | E_r) \geq P_\lambda(A_r) P_\lambda(D_3(r)). \tag{4.16}$$

$\square$

We let the cell size of the lattice $\mathbb{L}_n$ go to zero along a sequence $\{a_n\}_{n \geq 1}$ such that (4.15) holds for every $n \geq 1$. Define the $\mathbb{L}_n$-crossing probabilities as follows:

$\sigma_n^*((l_1, l_2), \lambda, 1)$

$\quad := P_\lambda\{$there exists a vacant L–R $\mathbb{L}_n$-crossing of $[0, l_1] \times [0, l_2]\}$,

$\sigma_n^*((l_1, l_2), \lambda, 2)$

$\quad := P_\lambda\{$there exists a vacant T–B $\mathbb{L}_n$-crossing of $[0, l_1] \times [0, l_2]\}$.

Now suppose (4.11) and (4.12) hold. For $i = 1, 2$ let $\{\delta_i(n)\}_{n \geq 1}$ be a sequence such that $\delta_i(n) \to \delta_i$ as $n \to \infty$ and $\sigma_n^*((l_1, l_2), \lambda, 1) \geq \delta_1(n)$ and $\sigma_n^*((l_3, l_2), \lambda, 2) \geq \delta_2(n)$. This is possible since $\sigma_n^*((l_1, l_2), \lambda, 1) \to \sigma^*((l_1, l_2), \lambda, 1)$ and $\sigma_n^*((l_3, l_2), \lambda, 2) \to \sigma^*((l_3, l_2), \lambda, 2)$ as $n \to \infty$. We will now provide a lower bound for $P_\lambda(D_3(r))$. Consider the event

$$\text{ref}(D_3(r)) := \{\text{there exists a vacant } \mathbb{L}_n\text{-path } s'' = (C_0'', \ldots, C_\mu'')$$

$$\text{such that } C_0 \cap ([l_1/4, 7l_1/4] \times \{l_2\}) \neq \emptyset,$$

$$C_\mu'' \cap (\text{ref}(F(r) \cup \Gamma_r)) \neq \emptyset, s'' \subseteq \overline{J_3^+(r)} \text{ and}$$

$$C_k'' \cap (([l_1/4, 7l_1/4] \times \{l_2\}) \cup (\mathrm{ref}(F(r) \cup \Gamma_r)))) = \emptyset$$
$$\text{for all } k = 1, \ldots, \mu - 1\}.$$

Clearly,

$$P_\lambda(D_3(r)) = P_\lambda(\mathrm{ref}(D_3(r))). \tag{4.17}$$

Also

$$
\begin{aligned}
P_\lambda(D_3(r) \cup (\mathrm{ref}(D_3(r)))) =\ & P_\lambda\{\exists \mathbb{L}_n\text{-path } \hat{s} = (\hat{C}_0, \ldots, \hat{C}_k) \text{ for a } k \geq 1 \\
& \text{such that } (\hat{C}_1, \ldots, \hat{C}_k) \subseteq J_3^+(r), \\
& \hat{C}_0 \cap ([l_1/4, 7l_1/4] \times \{l_2\}) \neq \emptyset \text{ and} \\
& \hat{C}_k \cap [(F(r) \cup \Gamma_r) \cup (\mathrm{ref}(F(r) \cup \Gamma_r))] \neq \emptyset\} \\
\geq\ & \sigma_n^*((l_3, l_2), \lambda, 2) \\
\geq\ & \delta_2(n),
\end{aligned}
$$

and hence, from (4.17), we have

$$P_\lambda(D_3(r)) \geq \delta_2(n)/2. \tag{4.18}$$

Now let $s' := (C_0', \ldots, C_\kappa')$ be a self-avoiding L–R $\mathbb{L}_n$-crossing of $[0, l_1 - R] \times [0, l_2]$ with $C_0' \cap (\{0\} \times [0, l_2]) \neq \emptyset$, $C_\kappa' \cap (\{l_1 - R\} \times [0, l_2]) \neq \emptyset$ and $(C_1', \ldots, C_{\kappa-1}') \subseteq (0, l_1 - R) \times [0, l_2]$. Let $C_{i_{l(s')}}'$ be, as defined earlier, the 'last intersection' of s' with $\{l_1/4\} \times [0, l_2]$. Let $Y(s')$ denote the second coordinate of the centre point of the cell $C_{i_{l(s')}}'$. We define the following events:

$H_1 :=$ {there exists a vacant L–R $\mathbb{L}_n$-crossing s of $[0, l_1 - R] \times [0, l_2]$
with $Y(s) \leq l_2/2$},

$H_2 :=$ {there exists a vacant L–R $\mathbb{L}_n$-crossing s of $[0, l_1 - R] \times [0, l_2]$
with $Y(s) \geq l_2/2$}.

Clearly, $P_\lambda(H_1) = P_\lambda(H_2)$ and $P_\lambda(H_1 \cup H_2) = P_\lambda\{$there exists a vacant L–R $\mathbb{L}_n$-crossing s of $[0, l_1 - R] \times [0, l_2]\}$. Thus $P_\lambda(H_1) \geq \delta_1(n)/2$. But, $H_1 \subseteq \{\mathcal{R} \text{ exists and } Y(\mathcal{R}) \leq l_2/2\}$, so

$$P_\lambda\{\mathcal{R} \text{ exists and } Y(\mathcal{R}) \leq l_2/2\} \geq \delta_1(n)/2. \tag{4.19}$$

4.5 RSW – proof

Now we are in a position to prove the RSW lemma.

Proof of the RSW lemma   First we show that

$P_\lambda\{$there exists a vacant L–R $\mathbb{L}_n$-crossing r' of $[0, l_1] \times [0, l_2]$ with

   $Y(r') \leq l_2/2$ and there exists a vacant $\mathbb{L}_n$-path s' with

   $s' \cap F'(r') \neq \emptyset, s' \subseteq \overline{J_3^+(r')}$ and $s' \cap ([l_1/4, 7l_1/4] \times \{l_2\}) \neq \emptyset\}$

   $\geq C_1(\lambda, R)\delta_1(n)\delta_2(n)/4,$ (4.20)

where $C_1(\lambda, R) = \exp\{-\lambda(3R + a_n)(2R + 2a_n)\}$. Here $F'(r')$ denotes the piece of r' after the last intersection of r' with the line $\{l_1/4\} \times [0, l_2]$ and $\overline{J_3^+(r')}$ is the closure of the region $J_3^+(r')$, which is defined as in the definition of $J_3^+(r)$ in Section 4.3 with $F'(r')$ instead of $F(r) \cup \Gamma_r$ and ref$(F'(r'))$ instead of ref$(F(r) \cup \Gamma_r)$.

To this end we observe that for an L–R $\mathbb{L}_n$-crossing r of $[0, l_1 - R] \times [0, l_2]$,

$$\{\mathcal{R} = r\} \cap A_r \subseteq \{\mathcal{R} = r, \Gamma_r \text{ is vacant}\}.   (4.21)$$

Indeed, if there are no balls centred in A_r' and no balls intersecting r then there cannot be any ball intersecting Γ_r. Moreover, for the L–R $\mathbb{L}_n$-crossing $r \cup \Gamma_r$ of $[0, l_1] \times [0, l_2]$, we have $F'(r \cup \Gamma_r) = F(r) \cup \Gamma_r$.

Thus from (ii) of Lemma 4.2, Lemma 4.3, (4.18), (4.19) and (4.21) we obtain

$P_\lambda\{$there exists a vacant L–R $\mathbb{L}_n$-crossing r' of $[0, l_1] \times [0, l_2]$ with

   $Y(r') \leq l_2/2$ and there exists a vacant $\mathbb{L}_n$-path s' with

   $s' \cap F'(r') \neq \emptyset, s' \subseteq \overline{J_3^+(r')}$ and $s' \cap ([l_1/4, 7l_1/4] \times \{l_2\}) \neq \emptyset\}$

$\geq P_\lambda\{\mathcal{R} \text{ exists}, Y(\mathcal{R}) \leq l_2/2, \Gamma_\mathcal{R} \text{ is vacant and } D_3(\mathcal{R}) \text{ occurs}\}$

$= P_\lambda\left(\bigcup_{\{r:Y(r)\leq l_2/2\}} \{\mathcal{R} = r, \Gamma_r \text{ is vacant and } D_0(r) \text{ occurs}\} \right)$

$= \sum_r P_\lambda\{\mathcal{R} = r, \Gamma_r \text{ is vacant and } D_0(r) \text{ occurs}\}$

$\geq \sum_r P_\lambda(\{\mathcal{R} = r\} \cap A_r \cap D_0(r))$

$= \sum_r P_\lambda(\{\mathcal{R} = r\})P_\lambda(A_r \cap D_0(r)|\{\mathcal{R} = r\})$

$\geq \sum_r P_\lambda(\{\mathcal{R} = r\})P_\lambda(A_r)P_\lambda(D_3(r))$

$\geq \sum_r \frac{1}{2}P_\lambda(\{\mathcal{R} = r\})\delta_2(n) \exp\{-\lambda(3R + a_n)(2R + 2a_n)\}$

$\geq \frac{1}{4}\delta_1(n)\delta_2(n) \exp\{-\lambda(3R + a_n)(2R + 2a_n)\}.$ (4.22)

(In the calculations above, $\cup_r$ and $\sum_r$ are, respectively, the union and sum over all L–R $\mathbb{L}_n$-crossings r of $[0, l_1 - R] \times [0, l_2]$ with $Y(r) \leq l_2/2$.) This proves (4.20).

Now, for any L–R crossing γ of $[0, l_1] \times [0, l_2]$, let $F'(\gamma)$ denote the piece of γ after its last intersection with the line $\{l_1/4\} \times [0, l_2]$. Also, let $Y(\gamma) :=$ $\inf\{y : (l_1/4, y) \in F'(\gamma) \cap \{l_1/4\} \times [0, l_2]\}$ and $J_3^+(\gamma)$ the closure of $J_3^+(\gamma) :=$ $\{(x, y) \in [l_1/4, 7l_1/4] \times [0, l_2] : (x, y)$ can be connected to $[l_1/4, 7l_1/4] \times \{l_2\}$ by a continuous curve $\tilde{\gamma}$ lying in $[l_1/4, 7l_1/4] \times [0, l_2]$ and $\tilde{\gamma} \cap (F'(\gamma) \cup$ ref$(F'(\gamma))) = \emptyset\}$, where ref$(F'(\gamma))$ is the reflection of $F'(\gamma)$ in $\{l_1\} \times [0, l_2]$. Now making the lattice finer and taking limits in (4.22) along the sequence $\{a_n\}$ chosen earlier, we obtain,

$$P_\lambda\{\text{there exists a vacant L–R crossing } \gamma_{1L} \text{ of } [0, l_1] \times [0, l_2] \text{ with}$$
$$Y(\gamma_{1L}) \leq l_2/2 \text{ and there exists a vacant path } \gamma_{1T} \text{ with}$$
$$\gamma_{1T} \cap F'(\gamma_{1T}) \neq \emptyset, \gamma_{1T} \subseteq \overline{J_3^+(\gamma_{1L})} \text{ and}$$
$$\gamma_{1T} \cap ([l_1/4, 7l_1/4] \times \{l_2\}) \neq \emptyset\}$$
$$\geq [\delta_1\delta_2 \exp(-6\lambda R^2)]/4. \tag{4.23}$$

An iterative procedure will now complete the proof. Suppose the event described in (4.23) occurs; then $\gamma_{1L} \cup \gamma_{1T}$ contains a vacant path ζ connecting $\{l_1/4\} \times [0, l_2/2]$ to $[l_1/4, 7l_1/4] \times \{l_2\}$. If for some $M > 0$, ζ contains any point in the region $[(l_1/4) + M, \infty) \times [0, l_2]$, then $\gamma_{1L} \cup \gamma_{1T}$ provides a vacant L–R crossing of $[0, (l_1/4) + M] \times [0, l_2]$. Otherwise, if $\zeta \subset ([0, (l_1/4) + M] \times [0, l_2])$, then consider a vacant L–R crossing γ_{2L} of $[l_1/4, (l_1/4) + M] \times [0, l_2]$ with $Y_2(\gamma_{2L}) \geq l_2/2$, where $Y_2(\gamma_{2L})$ is the second coordinate of the initial point γ_{2L}; i.e., $\gamma_{2L} \cap (\{l_1/4\} \times [0, l_2]) = (l_1/4, Y_2(\gamma_{2L}))$. Since ζ connects $\{l_1/4\} \times [0, l_2/2]$ to $[l_1/4, 7l_1/4] \times \{l_2\}$ and lies in $[0, (l_1/4) + M] \times [0, l_2]$, we have $\gamma_{2L} \cap \zeta \neq \emptyset$. Thus $\gamma_{1L} \cup \gamma_{1T} \cup \gamma_{2L}$ provides an L–R crossing of $[0, (l_1/4) + M] \times [0, l_2]$. Hence, applying the FKG inequality, we have

$$P_\lambda\{\text{there exists a vacant L–R crossing of } [0, (l_1/4) + M] \times [0, l_2]\}$$
$$\geq P_\lambda\{\text{the event in (4.23) occurs and there exists a vacant L–R}$$
$$\text{crossing } \gamma_{2L} \text{ of } [l_1/4, (l_1/4) + M] \times [0, l_2] \text{ with } Y_2(\gamma_{2L}) \geq l_2/2\}$$
$$\geq \delta_1\delta_2 \exp\{-6\lambda R^2\} P_\lambda(V_1)/4, \tag{4.24}$$

where $V_1 = \{$there exists a vacant L–R crossing γ_{2L} of $[l_1/4, (l_1/4) + M] \times [0, l_2]$ with $Y_2(\gamma_{2L}) \geq l_2/2\}$. Clearly

$$P_\lambda(V_1) \geq \sigma^*((M, l_2), \lambda, 1)/2, \tag{4.25}$$

and (4.24) and (4.25) yield

$$\sigma^*(((l_1/4) + M, l_2), \lambda, 1) \geq \delta_1 \delta_2 \exp\{-6\lambda R^2\} \sigma^*((M, l_2), \lambda, 1)/8. \quad (4.26)$$

Taking $M = M_0 := l_1$ in (4.26),

$$\sigma^*((5l_1/4, l_2), \lambda, 1) \geq \delta_1 \delta_2 \exp\{-6\lambda R^2\} \sigma^*((l_1, l_2), \lambda, 1)/8$$
$$\geq \delta_1^2 \delta_2 \exp\{-6\lambda R^2\}/8. \quad (4.27)$$

Again, from (4.26), taking $M = M_1 := M_0 + l_1/4$, we have

$$\sigma^*((3l_1/2, l_2), \lambda, 1) \geq \delta_1 \delta_2 \exp\{-6\lambda R^2\} \sigma^*((5l_1/4, l_2), \lambda, 1)/8,$$

which after applying (4.27) yields a lower bound on $\sigma^*((3l_1/2, l_2), \lambda, 1)$. Repeating this procedure, we obtain

$$\sigma^*((M_{j+1}, l_2), \lambda, 1) \geq \delta_1 \delta_2 \exp\{-6\lambda R^2\} \sigma^*((M_j, l_2), \lambda, 1)/8,$$

where $M_j = l_1 + \frac{j}{4}l_1$, for every $j \geq 1$. Thus, for every $k \geq 1$, we can recursively obtain the lower bound for $\sigma^*((kl_1, l_2), \lambda, 1)$. This completes the proof of the theorem. $\qquad \square$

4.6 Equality of the critical densities

In this section we show that, in two dimensions, the various critical densities defined in this chapter are all equal. Unfortunately, the higher dimensional analogue of this result is not known.

The vacancy structure in the continuum model corresponds to the 'dual' structure of the discrete percolation model on lattices. However, this correspondence is rather rough, in the sense that while the 'dual' structure of a discrete model can be defined without any reference to the original percolation model, the vacancy structure in the continuum arises as the complement of the occupancy structure. It is for this reason that analysing the 'dual' of a suitable discrete percolation model will not give us the equality of the critical densities defined through the vacancy structure as we were able to do in Theorem 3.4 for the critical densities defined through the occupancy structure.

The proof of the equality of the critical densities defined through the vacancy structure relies on the RSW theorem and as such the two-dimensional restriction of the RSW theorem carries through. However, a bonus of this proof is that we obtain a sharp transition in two dimensions; i.e., we obtain that $\lambda_c = \lambda_c^*$. Thus, in two dimensions, for $\lambda < \lambda_c$, we have a regime where there is no unbounded occupied component and there is at least one unbounded vacant component, while for $\lambda > \lambda_c$, no unbounded vacant components exist and there is at least one unbounded occupied component. Of course, this statement is probabilistic and holds only with probability 1. Together with the results from Section 3.6 and the results in the next section, we see that in either of these regimes there is almost surely exactly one unbounded component of the appropriate type. We show the following:

Theorem 4.3 *For a two-dimensional Poisson Boolean model* (X, ρ, λ) *with* ρ *bounded almost surely, we have* $\lambda_c^* = \lambda_T^* = \lambda_S^*$.

Theorem 4.4 *For a two-dimensional Poisson Boolean model* (X, ρ, λ) *with* ρ *bounded almost surely, we have* $\lambda_c = \lambda_c^*$.

Before we prove the theorems, we first state and prove a preliminary result. This lemma allows us to give a lower bound of the probability of the existence of an occupied crossing of a rectangle in terms of the probability of the existence of an occupied crossing of a bigger rectangle which is however of the same length in the direction of the crossing as that of the smaller rectangle.

Lemma 4.4 *Let n and k be positive integers and let* $\eta > 0$ *be such that*

$$\sigma^*((n, (1+2k)n), \lambda, 1) > \eta; \tag{4.28}$$

then, for any $t > 0$ *and for some* $f(t, k, \eta) > 0$,

$$\sigma^*((n, (1+2t)n), \lambda, 1) > f(t, k, \eta). \tag{4.29}$$

(The point here is that f does not depend on n.)

Proof If $t \geq k$ then the lemma holds trivially. Otherwise, we set $H_u := \{0\} \times [(1+k)n, (1+2k)n]$, $H_m := \{0\} \times [kn, (1+k)n]$ and $H_b := \{0\} \times [0, kn]$ and for $j = u, m$ or b we define the events $A_j := \{$there is an L–R vacant crossing s of the rectangle $[0, n] \times [0, (1 + 2k)n]$ with $s \cap H_j \neq \emptyset\}$. Clearly, $A_u \cup A_m \cup A_b = \{$there exists an L–R vacant crossing of $[0, n] \times [0, (1+2k)n]\}$. Moreover, since A_u, A_m and A_b are all decreasing events, applying the FKG inequality and noting that $P_\lambda(A_u) = P_\lambda(A_b)$, we have

$$1 - \sigma^*((n, (1+2k)n), \lambda, 1) = P_\lambda(A_u^c \cap A_m^c \cap A_b^c)$$
$$\geq P_\lambda(A_u^c) P_\lambda(A_m^c) P_\lambda(A_b^c)$$
$$= (1 - P_\lambda(A_m))(1 - P_\lambda(A_b))^2.$$

Thus at least one of the following (4.30) and (4.31) holds:

$$P_\lambda(A_m) \geq 1 - (1 - \sigma^*((n, (1+2k)n), \lambda, 1))^{1/3}$$
$$\geq 1 - (1 - \eta)^{1/3} := \eta' \text{ (say)}, \tag{4.30}$$

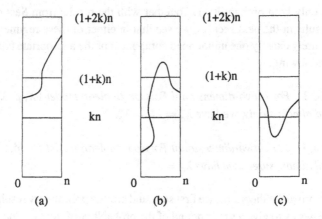

Figure 4.8. The three possibilities B_u, B_m and B_b.

$$P_\lambda(A_b) \geq 1 - (1 - \sigma^*((n, (1+2k)n), \lambda, 1))^{1/3}. \qquad (4.31)$$

Suppose first (4.30) holds. We observe that $A_m \subseteq B_u \cup B_m \cup B_b$, where

$B_u := \{$there exists a vacant L–R crossing of $[0, n] \times [kn, (1 + 2k)n]\}$,

$B_m := \{$there exists a vacant T–B crossing of $[0, n] \times [kn, (1 + k)n]\}$,

$B_b := \{$there exists a vacant L–R crossing of $[0, n] \times [0, (1 + k)n]\}$

(see Figure 4.8). Since B_u, B_m and B_b are all decreasing events, (4.30) and an application of the FKG inequality yields

$$1 - \eta' \geq P_\lambda(B_u^c \cap B_m^c \cap B_b^c)$$
$$\geq (1 - P_\lambda(B_m))(1 - P_\lambda(B_b))^2.$$

Thus at least one of the following (4.32) and (4.33) must hold:

$$P(B_m) \geq 1 - (1 - \eta')^{1/3}, \qquad (4.32)$$

$$P(B_b) \geq 1 - (1 - \eta')^{1/3}. \qquad (4.33)$$

Next, suppose that (4.31) holds. We observe that $A_b \subseteq B_m \cup B_b$ and so an application of the FKG inequality implies that at least one of (4.32) and (4.33) is true. In either case (i.e., (4.32) or (4.33)), the monotonicity of the crossing probabilities yields that

$$\sigma^*((n, (1 + k)n), \lambda, 1) \geq 1 - (1 - \eta')^{1/3}. \qquad (4.34)$$

Now take $k_0 := k$ in (4.34) and define $k_1 := k_0/2$. Then we have $\sigma^*((n, (1 + 2k_1)n), \lambda, 1) \geq 1 - (1 - \eta_0)^{1/3}$, where $\eta_0 := 1 - (1 - \sigma^*((n, (1 + 2k_0)n),$

$\lambda, 1))^{1/3}$. Defining successively $k_j := k_{j-1}/2$ for $j \geq 1$, we see that $\sigma^*((n, (1+2k_j)n), \lambda, 1) \geq 1 - (1 - \sigma^*((n, (1+2k_{j-1})n), \lambda, 1))^{1/3}$. Since $k_j \downarrow 0$ as $j \uparrow \infty$, this iterative procedure yields (4.29) for a suitable $f(t, k, \eta) > 0$. $\square$

Proof of Theorems 4.3 and 4.4   Throughout this proof we assume that

$$0 < \rho \leq R, \tag{4.35}$$

for some $R > 0$, and that the Poisson Boolean model is defined on $\mathbb{R}^2$. First we list some of the inequalities we obtain quite easily. It is clear that

$$\lambda_c^* \leq \lambda_D^*. \tag{4.36}$$

Moreover, by (4.4),

$$\lambda_D^* \leq \lambda_S^*. \tag{4.37}$$

In view of (4.36) and (4.37), to prove Theorem 4.3 and 4.4, it suffices to show that

$$\lambda_c \leq \lambda_c^*, \tag{4.38}$$

and

$$\lambda_S^* \leq \lambda_c. \tag{4.39}$$

To show (4.38), let $\lambda < \lambda_c$ and note that if instead of vacancy we consider occupancy in the proof of part (b) of Theorem 4.1 we obtain

$$\sum_{k=1}^{\infty} \sigma((3^k, 3^{k+1}), \lambda, 1) < \infty. \tag{4.40}$$

However, $\sigma((3^k, 3^{k+1}), \lambda, 1) + \sigma^*((3^k, 3^{k+1}), \lambda, 2) = 1$ because, if an occupied crossing does not exist in the horizontal direction then, almost surely, there must exist a vacant crossing in the vertical direction and vice versa. Hence, from (4.40) and the Borel–Cantelli lemma we have

$$P_\lambda\{\text{there is a vacant T–B crossing } t_k \text{ of } [0, 3^k] \times [0, 3^{k+1}]$$
$$\text{for all large } k\} = 1. \tag{4.41}$$

The rotation invariance allows us to restate (4.41) as

$$P_\lambda\{\text{there is a vacant L–R crossing } l_k \text{ of } [0, 3^{k+2}] \times [0, 3^{k+1}]$$
$$\text{for all large } k\} = 1. \tag{4.42}$$

Now a vertical crossing t_k of $[0, 3^k] \times [0, 3^{k+1}]$ and a horizontal crossing l_k of $[0, 3^{k+2}] \times [0, 3^{k+1}]$ must intersect. Also t_{k+1} and l_k must intersect. Thus the vacant crossings t_k and l_k defined in (4.41) and (4.42) combine to give an unbounded vacant component in the first quadrant. Hence, using the fact that vacant components are open, we obtain

$$P_\lambda\{d(V(x)) = \infty \text{ for some } x \in Q^2\} = 1,$$

where Q^2 is the set of all points in $\mathbb{R}^2$ with rational coordinates. Since

$$P_\lambda\{d(V(x)) = \infty \text{ for some } x \in Q^2\}$$
$$\leq \sum_{x \in Q^2} P_\lambda\{d(V(x)) = \infty\},$$

we have by translation invariance $P_\lambda\{d(V) = \infty\} > 0$. Thus $\lambda \leq \lambda_c^*$ and this completes the proof of (4.38).

Finally, to prove (4.39), we show that for $\lambda < \lambda_S^*$, there are infinitely many annuli around the origin, each annulus containing a vacant circuit with a probability larger than a positive constant. Thus a Borel–Cantelli argument would show that, with probability 1, there exist infinitely many circuits surrounding the origin and hence both $d(W)$ and $\ell(W)$ are finite. To this end we introduce another critical density:

$$\hat{\lambda}_S^* := \sup\{\lambda : \text{ there exists } 0 \leq n_1 \leq n_2 \cdots \text{ with } n_k \uparrow \infty \text{ as } k \uparrow \infty,$$

such that, for every $k \geq 1$ and for some $\delta > 0$, the following

hold:

(i) $5n_{2k-1} > 4n_{2k}$,

(ii) $\sigma^*((n_{2k-1}, n_{2k}), \lambda, 1) \geq \delta > 0$,

(iii) $\sigma^*((5n_{2k-1}/4, n_{2k}), \lambda, 2) \geq \delta > 0\}$.

We first show that

$$\lambda_S^* \leq \hat{\lambda}_S^*. \tag{4.43}$$

Let $\lambda < \lambda_S^*$. Then there exists an increasing sequence $\{m_k\}_{k \geq 1}$ of positive numbers, with $m_k \uparrow \infty$ as $k \uparrow \infty$, and some $\eta > 0$ such that $\sigma^*((m_k, 3m_k), \lambda, 1) > \eta$ for each $k \geq 1$. Now, for every $k \geq 1$, we take $n_{2k-1} = 5m_k/6$ and $n_{2k} = m_k$. Applying first the monotonicity property of the crossing probabilities and Lemma 4.4, we have $\sigma^*((n_{2k-1}, n_{2k}), \lambda, 1) > \delta$ and $\sigma^*((5n_{2k-1}/4, n_{2k}), \lambda, 2) > \delta$ for some $0 < \delta < \eta$ and for every $k \geq 1$. Thus $\lambda \leq \hat{\lambda}_S^*$ and consequently $\lambda_S^* \leq \hat{\lambda}_S^*$. This proves (4.43).

Next suppose that for some $\delta > 0$ and for every $k \geq 1$, n_{2k} and n_{2k-1} are such that (i), (ii) and (iii) in the definition of $\hat{\lambda}_S^*$ hold. For every $k \geq 1$, we put

$$m_k = n_{2k-1} \quad \text{and} \quad \tilde{m}_k = m_k + n_{2k}.$$

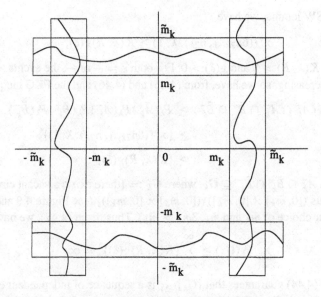

Figure 4.9. The events $\tilde{A}_k^+$, $\tilde{A}_k^-$, $\tilde{B}_k^+$ and $\tilde{B}_k^-$.

We also assume that the sequence $\{n_k\}_{k\geq 1}$ is chosen so that

$$m_{k+1} > 3m_k + 2R. \qquad (4.44)$$

Consider the rectangles

$$A_k^+ = [-3m_k, 3m_k] \times [m_k, \tilde{m}_k],$$

$$A_k^- = [-3m_k, 3m_k] \times [-\tilde{m}_k, -m_k],$$

$$B_k^+ = [m_k, \tilde{m}_k] \times [-3m_k, 3m_k],$$

$$B_k^- = [-\tilde{m}_k, -m_k] \times [-3m_k, 3m_k].$$

Also, for $\epsilon = +$ or $-$, we let

$$\tilde{A}_k^\epsilon = \{\text{there exists an L–R vacant crossing of } A_k^\epsilon\}$$

and

$$\tilde{B}_k^\epsilon = \{\text{there exists a T–B vacant crossing of } B_k^\epsilon\}.$$

For an illustration of these events, see Figure 4.9. By the invariance properties of the model,

$$P_\lambda(\tilde{A}_k^+) = P_\lambda(\tilde{A}_k^-) = P_\lambda(\tilde{B}_k^+) = P_\lambda(\tilde{B}_k^-) = \sigma^*((6n_{2k-1}, n_{2k}), \lambda, 1). \qquad (4.45)$$

By the RSW lemma, we have

$$\sigma^*((6n_{2k-1}, n_{2k}), \lambda, 1) \geq K(\lambda, R)g(\delta), \qquad (4.46)$$

for some $K(\lambda, R) > 0$ and $g(\delta) > 0$. For both $\epsilon = +$ or $-$, the events $\tilde{A}_k^\epsilon$ and $\tilde{A}_k^\epsilon$ are decreasing, so we have, from (4.45) and (4.46) and the FKG inequality,

$$
\begin{aligned}
P_\lambda(\tilde{A}_k^+ \cap \tilde{A}_k^- \cap \tilde{B}_k^+ \cap \tilde{B}_k^-) &\geq P_\lambda(\tilde{A}_k^+)P_\lambda(\tilde{A}_k^-)P_\lambda(\tilde{B}_k^+)P_\lambda(\tilde{B}_k^-) \\
&\geq \{\sigma^*((6n_{2k-1}, n_{2k}), \lambda, 1)\}^4 \\
&\geq \{K(\lambda, R)g(\delta)\}^4. \qquad (4.47)
\end{aligned}
$$

But $\tilde{A}_k^+ \cap \tilde{A}_k^- \cap \tilde{B}_k^+ \cap \tilde{B}_k^- \subseteq G_k$, where $G_k := \{$there exists a vacant circuit in the annulus $([0, \tilde{m}_k] \times [0, \tilde{m}_k]) \backslash ([0, m_k] \times [0, m_k])\}$ (see Figure 4.9 and note that by our choice of m_k and $\tilde{m}_k$, $3m_k > \tilde{m}_k$). Thus, from (4.47), we have

$$\sum_{k=1}^\infty P_\lambda(G_k) \geq \sum_{k=1}^\infty \{K(\lambda, R)g(\delta)\}^4 = \infty.$$

However, (4.44) guarantees that $\{G_k\}_{k\geq 1}$ is a sequence of independent events, so that the Borel–Cantelli lemma yields that $P_\lambda\{G_k$ occurs infinitely often$\} = 1$. But $P_\lambda\{$there are infinitely many vacant circuits around the origin$\} \geq P_\lambda\{G_k$ occurs infinitely often$\} = 1$, whence

$$P_\lambda\{d(W) = \infty\} = 0. \qquad (4.48)$$

Now, if $\lambda < \hat{\lambda}_S^*$, then there exists a $\delta > 0$ and a sequence $\{n_k\}_{k\geq 1}$ satisfying (i), (ii) and (iii) in the definition of $\hat{\lambda}_S^*$ and consequently, from (4.48), $\lambda \leq \lambda_c$. This, along with (4.43), proves (4.39) and completes the proof of Theorems 4.3 and 4.4. □

An immediate consequence of this theorem is that, in two dimensions, for $\lambda > \lambda_c$, $\sigma^*((n, 3n), \lambda, 1) \to 0$ as $n \to \infty$ and thus $\sigma((n, 3n), \lambda, 2) \to 1$ as $n \to \infty$, while for $\lambda < \lambda_c$, $\sigma((n, 3n), \lambda, 1) \to 0$ as $n \to \infty$ and thus $\sigma^*((n, 3n), \lambda, 1) \to 1$ as $n \to \infty$. This argument can be easily generalised along the lines of the proof of Lemma 4.4 to yield

Corollary 4.1 *Consider a Poisson Boolean model (X, ρ, λ) in two dimensions with ρ bounded. For $\lambda > \lambda_c$, we have $\sigma((kn, n), \lambda, 1) \to 1$ as $n \to \infty$ for every $k \geq 1$, and for $\lambda < \lambda_c$, we have $\sigma^*((kn, n), \lambda, 1) \to 1$ as $n \to \infty$ for every $k \geq 1$.*

Theorem 4.5 *Consider the Poisson Boolean model (X, ρ, λ) on $\mathbb{R}^d$ with $0 < \rho \leq R$ for some $R > 0$. For $d = 2$, we have $P_{\lambda_c}\{d(W) = \infty\} = 0$. However, for any $d \geq 2$, $E_{\lambda_c}(d(W)) = \infty$.*

Proof Suppose $P_{\lambda_c}\{d(W) = \infty\} > 0$. We shall show that this implies that

$$\sigma^*((n, 3n), \lambda_c, 1) \to 0 \text{ as } n \to \infty. \qquad (4.49)$$

Indeed, if (4.49) does not hold, then there exist $\delta > 0$ and an infinite sequence $n_1, n_2, \ldots$ such that, for all $i \geq 1$, $\sigma^*((n_i, 3n_i), \lambda_c, 1) \geq \delta$. Without loss of generality we may assume that, for all $i \geq 1$,

$$6n_{i+1} \geq 9n_i + 2R. \qquad (4.50)$$

The RSW lemma implies that there exist $0 < \delta_0 < \delta$ such that for all $i \geq 1$,

$$\sigma^*((18n_i, 3n_i), \lambda_c, 1) \geq \delta_0. \qquad (4.51)$$

As in the proof of the previous theorem, for every $i \geq 1$, the FKG inequality together with (4.51) allows us to construct vacant circuits in the annulus $B_{18n_i} \setminus B_{15n_i}$ with a probability larger than δ_0^4. Thus, for $E_i := \{$there exists a vacant circuit in the annulus $B_{18n_i} \setminus B_{15n_i}\}$, we have

$$\sum_{i=1}^{\infty} P_{\lambda_c}(E_i) = \infty.$$

By our choice (4.50) of n_i, $E_1, E_2, \ldots$ is a sequence of independent events; thus an application of the Borel–Cantelli lemma yields

$$P_{\lambda_c}\{E_i \text{ occurs infinitely often}\} = 1,$$

and it follows that $P_{\lambda_c}\{d(W) = \infty\} = 0$. This contradiction establishes (4.49).

To complete the proof of the first part of the theorem, note that (4.49) implies that, for some $N \geq 0$,

$$\sigma^*((N, 3N), \lambda_c, 1) < \kappa_0, \qquad (4.52)$$

where κ_0 is as in Lemma 4.1. As in Chapter 3, since the event $\{$there is a vacant L–R crossing of the rectangle $[0, N] \times [0, 3N]\}$ depends on the bounded rectangle $[-R, N+R] \times [-R, 3N+R]$, $\sigma^*((N, 3N), \lambda, 1)$ is a continuous function of λ. Thus from (4.52) we have that for some $\lambda < \lambda_c$,

$$\sigma^*((N, 3N), \lambda, 1) < \kappa_0. \qquad (4.53)$$

However, (4.53) implies from Lemma 4.1 that, for this λ, $P_\lambda\{d(V) = \infty\} = 0$. This contradicts the fact that $\lambda_c = \lambda_c^*$ and thus proves the first part of the theorem.

To show the second part of the theorem, observe that if $E_{\lambda_c}(d(W)) < \infty$, then from Theorem 2.4, we have that

$$P_{\lambda_c}\{d(W(B_1)) \geq a\} \leq C_1 e^{-aC_2} \qquad (4.54)$$

for some positive constants C_1 and C_2. Now let $D_k = k + B_1$ for $k \in \mathbf{Z}^d$. Clearly, for the existence of an occupied crossing in the shorter direction of the rectangle $[0, n] \times [0, 3n] \times \cdots \times [0, 3n]$, there must be an occupied component of diameter at least n from some D_k with $k = (k_1, \ldots, k_d)$, $k_1 = 0$ and $0 \le k_i \le 3n$. Hence,

$$\sigma((n, 3n, 3n, \ldots, 3n), \lambda_c, 1)$$

$$\le P_{\lambda_c}\{d(W(D_k)) \ge n \text{ for some } k = (k_1, \ldots k_d)$$

$$\text{with } k_1 = 0 \text{ and } 0 \le k_i \le 3n\}$$

$$\le (3m)^{d-1} C_1 e^{-nC_2}$$

$$\to 0 \text{ as } n \to \infty \qquad (4.55)$$

where the last inequality follows from (4.54). For n sufficiently large, (4.55) yields that

$$\sigma((n, 3n, 3n, \ldots, 3n), \lambda_c, 1) < \kappa_0. \qquad (4.56)$$

A continuity argument as in the previous part yields, for some $\lambda > \lambda_c$,

$$\sigma((n, 3n, 3n, \ldots, 3n), \lambda, 1) < \kappa_0.$$

This along with Lemma 3.3 shows that for this λ, $P_\lambda\{d(W) = \infty\} = 0$, which contradicts the fact that $\lambda > \lambda_c$. This completes the proof of the theorem. □

4.7 Uniqueness

In Section 3.6, it was shown that in a Poisson Boolean model, there can be at most one unbounded occupied component. This section is devoted to the analogous result for the vacant region. As in the occupancy case, the result holds in its most general form:

Theorem 4.6   *In a Poisson Boolean model (X, ρ, λ), there can be at most one unbounded vacant component a.s.*

According to Proposition 3.1 we can with no loss of generality assume in this section that

$$E\rho^d < \infty. \qquad (4.57)$$

The idea of the proof is similar to that in Section 3.6. There is, however, one extra difficulty. In the proof for occupancy, we used the fact that different occupied components contain different points of the process and that the density of such points is finite. We can not use such a direct argument here. We have to find other objects which have finite density and which can not be 'shared' by different vacant components. To this end, we state the following geometric result:

Lemma 4.5 *If k d-dimensional balls intersect the unit cube $[0, 1]^d$ then the vacant region inside the unit cube has at most $c_d k^d$ connected components, where c_d is a constant which depends only on the dimension.*

Proof For ease of exposition, we first give the proof for the case $d = 3$. Without loss of generality we can assume that no ball is contained in the union of the others, since adding this particular ball would not affect the number of vacant components. So we consider three balls, none of which is contained in the union of the other two. We claim that the intersection of the boundaries of these balls consist of at most two points. To see this, note that if the intersection of the boundaries of the first two balls is not empty, it is the boundary γ of some circle. Denote the plane containing this circle by H. The intersection of all three boundaries can be larger than two points only if the intersection of the third ball with H is γ. It is easy to check that in this case there is a ball which is contained in the other two, a contradiction. We call a point in the intersection of three boundaries a *triple point*. Now each vacant component in $[0, 1]^d$ which does not intersect the boundary of the cube contains at least one triple point on its boundary, and a triple point can belong to only one vacant component. Hence there are at most $2\binom{k}{3}$ vacant components which do not intersect the boundary of the unit cube. Next, we look at the intersection of vacant components with the faces of the cube. The intersection of a three-dimensional ball with a face is a two-dimensional ball. Any vacant component which intersects a face of the unit cube but does not intersect any of its edges must contain at least one point of intersection of two such two-dimensional balls. For each face therefore there can be at most $2\binom{k}{2}$ such vacant components, giving a total number of at most $12\binom{k}{2}$ for all the faces of the cube. Finally, an analogous argument shows that the number of vacant components which intersect an edge is at most $12(k + 1)$. Hence, the total number of vacant components is bounded by $c_3 k^3$ for a suitable constant c_3. The argument in higher dimensions is essentially the same and is omitted. □

A random variable which has a Poisson distribution has finite moments of all orders. Combining Lemma 3.1, Lemma 4.5 and the fact that the box B_n is contained in the ball $S(0, n\sqrt{d})$, we conclude:

Proposition 4.1 *In a Poisson Boolean model, the number of connected components in $V \cap B_n$ has finite expectation for all n.*

As in the occupancy case, we first show that the number of unbounded vacant components can be only zero, one or infinity:

Proposition 4.2   *In a Poisson Boolean model, the number of unbounded vacant components equals either zero, one or infinity a.s.*

Proof   Using ergodicity as before, we see that the number of unbounded vacant components is an a.s. constant. Suppose that this number is equal to $K \geq 2$. For any region $A \subset \mathbb{R}^d$, we denote by $V[A]$ the vacant region which we obtain after we remove all points (and associated balls) in the complement A^c of A. Using the assumption that the model admits K unbounded vacant components, Lemma 3.1, and the obvious fact that the number of vacant unbounded components can not increase by removing finitely many balls, we conclude that for all n, $V[B_n^c]$ contains no more that K unbounded vacant components a.s. For all positive integers n, let E_n be the event that all unbounded vacant components in $V[B_n^c]$ have non-empty intersection with B_n. For n large enough, we have $P(E_n) > 0$. Since only finitely many balls intersect B_n, there are integers L_1 and L_2 such that the event

$$E_n \ \cap \ \{\text{there are at most } L_1 \text{ balls which intersect}$$
$$B_n \text{ and all these balls have radius at most } L_2\}$$

has positive probability. Now consider the annulus $B_{n+L_2} \setminus B_n$. We partition this annulus using the integer lattice and let $\mathcal{C}$ be the (finite) collection of cells in this lattice. Since $\mathcal{C}$ is finite, we can choose non-random cells $W_1, \ldots, W_{L_1}$ in $\mathcal{C}$ such that for $W := W_1 \cup \cdots \cup W_{L_1}$ the event

$$F_n := \{\text{all unbounded vacant components in } V[(B_n \cup W)^c]$$
$$\text{have non-empty intersection with } B_n \text{ and all balls with}$$
$$\text{non-empty intersection with } B_n \text{ are centred in } B_n \cup W\}$$

has positive probability. Note that F_n depends only on the points of the point process outside $B_n \cup W$ and their associated balls. Hence,

$$P(F_n \cap \{X(B_n \cup W) = 0\}) = P(F_n)P(X(B_n \cup W) = 0) > 0.$$

But if $F_n \cap \{X(B_n \cup W) = 0\}$ occurs, then there is only one unbounded vacant component. This is the desired contradiction and completes the proof.   □

Proof of Theorem 4.6   The idea of the proof is similar to that in the proof of Theorem 3.6. However, the application of Lemma 3.2 is much easier here for reasons which will be explained below.

According to Proposition 4.2 we need to rule out the possibility of having infinitely many unbounded vacant components. As before, we proceed by assuming that there are infinitely many such components and we want to derive a contradiction. It follows from the proof of Proposition 4.2 that we can find boxes $B_n \subset B_m$ such that with positive probability η, say, B_n is completely contained in a vacant unbounded component V' such that $V' \cap B_m^c$ contains at least three disjoint unbounded components. If this is the case, we call B_m an *encounter box*, and B_n its *central box*. The three unbounded components of $V' \cap B_m^c$ are again called *branches*. Translating this event over the vector $2mz$, for $z \in \mathbb{Z}^d$, gives the requirement for B_m^{2mz} to be an encounter box. For all L, the expected number of encounter boxes of the form B_m^{2mz} contained in the box B_{mL} is equal to ηL^d. Now any branch of an encounter box in B_{mL} intersects $A_L := B_{m(L+1)} \setminus B_{mL}$ and hence contains at least one component of $C^c \cap B_m^{2mz}$ for some $z \in \mathbb{Z}^d$ such that $B_m^{2mz} \subset A_L$. (Recall that C^c is the vacant region in space.) We choose one such component for each branch b and call it V_b. Now we want to apply Lemma 3.2. For this, let R be the set of all central boxes corresponding to encounter boxes in B_{mL}. For $r \in R$, the sets $C_r^{(i)}$ are defined as follows. Choose a branch b of the encounter box to which r belongs. Take all central boxes of other encounter boxes which are contained in b, together with V_b which is chosen before. Together these elements form one of the sets $C_r^{(i)}$. It is obvious that $\mathrm{card}(C_r^{(i)}) \geq 1$ for all r and i. Let S consist of all central boxes and the union of all sets $C_r^{(i)}$, $r \in R$. Applying Lemma 3.2 with $K = 1$ then yields the conclusion that $\mathrm{card}(S) \geq 2\,\mathrm{card}(R) + 2$. Hence the number of components V_b must be at least $\mathrm{card}(R) + 2$ since S consists of the union of these components together with $\mathrm{card}(R)$ other elements. Thus the expected number of vacant components of the form $C^c \cap B_m^{2mz}$ contained in A_L is at least ηL^d. However, it follows from Proposition 4.1 that this expected number can be at most cL^{d-1} for some constant $c > 0$, and this gives the desired contradiction for L sufficiently large. $\qquad\square$

REMARK: It is instructive to compare this proof to the proof of Theorem 3.6. The reason that the application of Lemma 3.2 is so much easier here is the fact that in our definition, a branch in the occupancy case need not contain points in the annulus A_L. For a vacant branch b, the component V_b defined above has to exist. This means that the volume-boundary argument is much easier in the latter case.

4.8 Continuity of the percolation function

In this section we investigate to what extent we can prove analogues of Theorem 3.9 and Theorem 3.7 for vacancy. For this, we define the vacant percolation function $\theta_\rho^*(\lambda)$ as the probability that in (X, ρ, λ) the vacant component of the origin is unbounded. The corresponding critical density is $\lambda_c^*(\rho)$, which can now be written as $\lambda_c^*(\rho) = \sup\{\lambda : \theta_\rho^*(\lambda) > 0\}$.

It was shown in Theorem 4.4 that whenever ρ is bounded almost surely and the dimension is 2, $\lambda_c^*(\rho) = \lambda_c(\rho)$. In conjunction with Theorem 3.7 this trivially gives that in dimension 2, $\lambda_c^*(\rho_k) \to \lambda_c^*(\rho)$ whenever all radii are uniformly bounded and $\rho_k \Rightarrow \rho$. Note, however, that in the proof of Theorem 3.7 we used the fact that $\lambda_S = \lambda_c$. The corresponding statement for vacancy is only known in two dimensions (Theorem 4.3).

What we can show in full generality is the continuity of the vacant percolation function as a function of λ, at least for $\lambda \neq \lambda_c^*(\rho)$:

Theorem 4.7 *In a Poisson Boolean model (X, ρ, λ), the vacant percolation function θ_ρ^* is a continuous function of λ for all $\lambda \neq \lambda_c^*(\rho)$.*

Proof As in the proof of Theorem 3.9, $\{d(V) = \infty\}$ is the decreasing limit of the events $F_n := \{0 \overset{v}{\leadsto} \partial(B_n)\}$ and $P_\lambda(F_n)$ is continuous in λ. Hence θ_ρ^* is the decreasing limit of a sequence of non-increasing continuous functions and is therefore continuous from the left.

For continuity from the right we fix $\lambda_0 < \lambda_c^*$. As in the proof of Theorem 3.9 we couple all processes $(X, \alpha\rho, \lambda_0)$ on the same probability space, this time for all $\alpha \geq 1$. Writing V_α for the vacant component of the origin in $(X, \alpha\rho, \lambda_0)$, we need to show that

$$P(d(V_1) = \infty, d(V_\alpha) < \infty \text{ for all } \alpha > 1) = 0.$$

In words, if $d(V_1)$ is infinite, then there is an $\alpha > 1$ such that if we multiply all radii by α, the vacant component of the origin remains unbounded. From scaling as in the proof of Theorem 3.9 we observe that for $\alpha > 1$ small enough, $\theta_{\alpha\rho}^*(\lambda_0) > 0$ and for this choice of α, there exists exactly one unbounded vacant component U_α, say (Theorem 4.6). It follows that U_α is contained in V_1. This implies that there exists a bounded, closed curve $\gamma \subset V_1$ connecting the origin to U_α. At this point, the argument is different from the occupancy case. From Lemma 3.1 and the fact that $E((\alpha\rho)^d) < \infty$ by assumption, we see that there are a.s. only finitely many balls in $(X, \alpha\rho, \lambda_0)$ which intersect γ and none of these balls intersect γ when the radius is reduced to the original value. It follows that for some value of $\alpha > 1$, no ball in $(X, \alpha\rho, \lambda_0)$ intersects γ and the proof is complete. $\square$

4.9 Notes

The first seven sections of this chapter are from Roy (1990), the uniqueness result of Section 4.8 is from Meester and Roy (1994), and the continuity result of Section 4.9 is from Sarkar (1995). The RSW lemma is an analogue of the RSW lemma of discrete percolation obtained by Russo (1978) and Seymour and Welsh (1978). One of the first applications of this result in discrete percolation was by Kesten (1980) to show $p_c = \frac{1}{2}$. Subsequently, it has been used to obtain a variety of results in two-dimensional discrete percolation. One of the main open questions in percolation theory is to obtain a higher-dimensional analogue of the RSW lemma. The difficulty in this arises from the fact that a crossing in one direction of a cube may not intersect a crossing in another direction of the cube. Although, in three dimensions, if we consider a sheet crossing in one direction and another sheet crossing in another direction, then they will intersect. A suitable higher-dimensional analogue of this may also be formulated. The RSW theorem for this will go through; however, for the full import of the result in higher dimensions, we need line crossings.

In the discrete case, the notion of the lowest crossing was introduced and conditioned on the lowest crossing; the configuration 'above' the crossing and the configuration 'below' the crossing are independent. In the continuum case, this independence does not exist and, as such, the term $C_k(\lambda, R)$ enters the picture. The independence structure of discrete percolation allows one to formulate a stronger RSW lemma for the discrete than the continuum version we have here. In particular, in the discrete percolation setup, it may be shown that if the crossing probability of a smaller rectangle tends to 1 then the crossing probability of the larger rectangle also tends to 1. Here, however, the term $C_k(\lambda, R)$ prevents such a strong result.

More recently, Alexander (1994) has obtained an RSW lemma for occupied crossings with fixed sized balls. The proof of this result also proceeds through a lowest crossing argument. The difficulty here lies in the definition of a lowest occupied crossing. Alexander has an ingenious way of defining the lowest occupied crossing to obtain enough conditional independence between the region above and the region below the conditioned lowest crossing. The steps of this proof are significantly different from that of the RSW lemma for vacant crossings given in this chapter. The dependency structure persists here too and this brings in a term $C_k(\lambda, R)$ in the formulation of the RSW result in the occupied case. Thus the statement of the result is quite similar to the RSW lemma for vacant crossings, the only difference being that σ^* has to be replaced by σ. This theorem will also provide the vacant counterpart of the first part of Theorem 4.5 to yield that $P_{\lambda_c^*}\{d(V) = \infty\} = 0$ in two dimensions. However, the second part of this theorem needs a vacancy version of Theorem 2.4 which is not known to be true.

5

Distinguishing features of the Poisson Boolean model

The title of this chapter needs some explanation. The word 'distinguishing' refers to two facts. In the first place, we are going to describe some fundamental differences between Boolean models with balls of a fixed radius and balls with random radii. These differences have to do with the so-called covered volume fraction and with the phenomena of compression and rarefaction. These notions will be introduced shortly. In the second place, the phenomena described in this chapter do not have natural analogues in the discrete setting.

5.1 The covered volume fraction

Loosely speaking, the covered volume fraction (CVF) is supposed to be the 'fraction' of space which is covered by balls. In order to give a more precise definition, let us first recall from (3.3) that the probability that the origin is covered in the Poisson Boolean model (X, ρ, λ) is equal to $1 - e^{-\lambda \pi_d E \rho^d}$. It then follows easily from Fubini's theorem that the expected Lebesgue measure of the occupied region in the unit cube is equal to the same number. The ergodic theorem now guarantees that if B_n is the box $[-n, n]^d$ as usual, then the limit

$$\lim_{n \to \infty} \frac{1}{(2n)^d} \ell(B_n \cap C) \tag{5.1}$$

exists a.s. and equals $1 - e^{-\lambda \pi_d E \rho^d}$. The limit in (5.1) is taken as the definition of the CVF:

Definition 5.1 *The covered volume fraction of a Poisson Boolean model* (X, ρ, λ) *is defined as the almost sure limit*

$$\lim_{n \to \infty} \frac{1}{(2n)^d} \ell(B_n \cap C)$$

and is therefore equal to $1 - e^{-\lambda \pi_d E \rho^d}$.

122

We remark that in case $E\rho^d = \infty$, the CVF is equal to 1.

Definition 5.2 *The* critical CVF *of a Poisson Boolean model* (X, ρ, λ) *is defined as*

$$A_c(\rho) := 1 - e^{-\lambda_c(\rho)\pi_d E\rho^d}.$$

Thus $A_c(\rho)$ is the fraction of space covered at criticality. Now consider two Boolean models, one with balls of a fixed radius r_1 and the other with balls of a fixed radius r_2, and denote their critical CVF by $A_c(r_1)$ and $A_c(r_2)$ respectively. It follows immediately from Definition 5.2 and Proposition 2.10 that

$$A_c(r_1) = A_c(r_2).$$

Hence in any Poisson Boolean model with balls of a fixed radius, the fraction of space covered at criticality is a constant depending only on the dimension and which we denote by $A_c = A_c^{(d)}$. A natural quantity to investigate is the critical CVF if the balls do not have a fixed radius. We shall show that the critical CVF is certainly not a universal constant among all possible Boolean models.

Theorem 5.1 *Consider Boolean models* (X, ρ, λ) *in* $\mathbb{R}^d$. *There exists a random variable* ρ *taking values* $a > 0$ *and* $b > 0$ *with probability* p *and* $1 - p$ *respectively, where* $a \neq b$, $0 < p < 1$ *such that*

$$A_c(\rho) > A_c^{(d)}. \tag{5.2}$$

Proof For ease of notation we give the proof for the case $d = 2$ and we write $A_c = A_c^{(2)}$. The higher-dimensional case uses the same argument, but the notation becomes rather tedious to handle. Let $0 < r_1 < r_2 < \infty$ be arbitrary positive numbers. Fix $\epsilon, \delta > 0$ such that

$$(2 - \epsilon - \delta)A_c - (1 - \epsilon)(1 - \delta)A_c^2 > A_c. \tag{5.3}$$

The expression in (5.3) will become clear in a moment. Next we choose $\lambda_2 < \lambda_c(r_2)$ such that the CVF of (X, r_2, λ_2) is equal to $(1 - \epsilon)A_c$. Also choose $\lambda_1 < \lambda_c(r_1)$ such that the CVF of (X, r_1, λ_1) is equal to $(1 - \delta)A_c$. Note that both processes are subcritical. Next we consider the superposition of these processes. We claim that the CVF of this superposition is strictly larger than A_c. To see this, note that the probability that the origin is covered in the superposition of the two processes is just the left-hand side of (5.3) and the claim follows.

Now consider the process (X, r_1, λ_1) and scale it by a factor $\alpha < 1$ to obtain a process which is equivalent in law to $(X, \alpha r_1, \alpha^{-2}\lambda_1)$. In other words, if (X, r_1, λ_1) consists of the points $\{x_1, x_2, \ldots\}$, with associated balls of radius

r_1, then the scaled model consists of the points $\{\alpha x_1, \alpha x_2, \ldots\}$ with associated balls of radius αr_1. (Note that in this way, we couple all processes together for $\alpha < 1$.) The CVF of $(X, \alpha r_1, \alpha^{-2}\lambda_1)$ does not depend on α, whence it follows from (5.3) that the CVF of the superposition of (X, r_2, λ_2) and $(X, \alpha r_1, \alpha^{-2}\lambda_1)$ is strictly larger than A_c. Our goal now is to show that this superposition is subcritical for α sufficiently small.

Fix a $0 < \kappa < \kappa_0$ where κ_0 is as in Lemma 3.3. Since $\lambda_2 < \lambda_c(r_2)$, Theorem 3.5 implies that $\lambda_2 < \lambda_S(r_2)$ and we can thus find a number N so large that

$$\sigma((N, 3N), \lambda_2, 1) < \tfrac{1}{3}\kappa.$$

If there is no occupied L–R crossing in $[0, N] \times [0, 3N]$, then there is a vacant T–B crossing defined in the obvious way. In other words, there is at least one component in $([0, N] \times [0, 3N]) \cap V$ intersecting the top and bottom sides of the rectangle. We can order these components from left to right, say, and the leftmost component is called L. Only finitely many balls intersect $[0, N] \times [0, 3N]$ a.s. and hence the boundary ∂L of L has only finitely many components a.s. Hence, for n large enough, the event $E_n := \{L$ exists and all components of $\partial L \cap \text{int}([0, N] \times [0, 3N])$ have distance at least n^{-1} from each other$\}$ has probability at least $1 - \tfrac{1}{2}\kappa$. We fix n_0 such that

$$P_{(\lambda_2, r_2)}(E_{n_0}) > 1 - \tfrac{1}{2}\kappa. \tag{5.4}$$

Next we turn again to (X, r_1, λ_1). Since $\lambda_1 < \lambda_c(r_1)$, it follows from Lemma 3.3 and the application of the FKG-inequality (Example 2.1) that for $B_1 = [-1, 1]^2$,

$$P_{(\lambda_1, r_1)}(d(W(B_1)) \geq b) \leq C_3 e^{-C_4 b},$$

for all $b > 0$, where C_3 and C_4 are again positive constants independent of b. Scaling down by a factor $\alpha < 1$ yields

$$P_{(\alpha^{-2}\lambda_1, \alpha r_1)}(d(W(B_\alpha)) \geq \alpha b) \leq C_3 e^{-C_4 b},$$

where $B_\alpha = [-\alpha, \alpha]^2$. Taking $\alpha = m^{-1}$ for some large integer m, and $b = (2\alpha n_0)^{-1}$ (with n_0 as in (5.4)), we obtain

$$P_{(m^2\lambda_1, m^{-1}r_1)}(d(W(B_{m^{-1}})) \geq (2n_0)^{-1}) \leq C_3 e^{-C_4 m/2n_0}. \tag{5.5}$$

Now we combine the conclusions obtained in (5.4) and (5.5). Divide $[0, N] \times [0, 3N]$ into $3N^2 m^2$ boxes with side length m^{-1}, and denote these boxes by $B^1, B^2, \ldots, B^{3N^2 m^2}$. Then, from (5.5), the probability that in the model $(X, m^{-1}r_1, m^2\lambda_1)$ the event

$$F_{n_0}^m := \bigcup_{i=1}^{3N^2 m^2} \{d(W(B^i)) \geq (2n_0)^{-1}\}$$

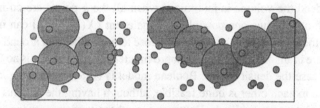

Figure 5.1. The superposition of the two processes. The small balls do not cross the gap left by the big balls. Hence no L–R crossing of the rectangle exists.

occurs has probability at most $3N^2 m^2 C_3 e^{-C_4 m/2n_0}$, which tends to zero for $m \to \infty$. We now fix an m_0 such that this probability is at most $\frac{1}{3}\kappa$. If E_{n_0} occurs in (X, r_2, λ_2) and $F_{n_0}^{m_0}$ does not occur in $(X, m_0^{-1} r_1, m_0^2 \lambda_1)$, then it follows that there is no occupied L–R crossing in $[0, N] \times [0, 3N]$ in the superposition of the two processes; see Figure 5.1. This superposition is in fact the model $(X, \rho, \lambda_2 + m_0^2 \lambda_1)$, where ρ is a random variable taking values r_2 and $m_0^{-1} r_1$ with probability $\lambda_2 (m_0^2 \lambda_1 + \lambda_2)^{-1}$ and $m_0^2 \lambda_1 (m_0^2 \lambda_1 + \lambda_2)^{-1}$, respectively. Hence, the probability of an occupied L–R crossing of $[0, N] \times [0, 3N]$ in $(X, \rho, \lambda_2 + m_0^2 \lambda_1)$ is at most $\frac{1}{2}\kappa + \frac{1}{3}\kappa < \kappa$. According to Lemma 3.3, this implies that this model is subcritical and this proves the theorem. $\qquad\square$

The following continuity result follows immediately from Definition 5.2 and Theorem 3.7. Together with Theorem 5.1 it also provides us with a class of radius distributions for which the strict inequality in (5.2) holds.

Theorem 5.2 *Let ρ_k and ρ be random variables such that for some $R > 0$ we have $0 \leq \rho \leq R$ and $0 \leq \rho_k \leq R$ a.s. for all $k \geq 1$. If $\rho_k \Rightarrow \rho$ then $A_c^d(\rho_k) \to A_c^d(\rho)$.*

5.2 Compression

In this section we investigate the structure of bounded components in a high-density Poisson Boolean model. Clearly, for any fixed $k \geq 0$, $P_\lambda\{X(W) = k\}$ is very small for large λ. Moreover, in the case when the balls are all of a fixed-size r, for the occurrence of the event $\{X(W) = k\}$ the component obtained by the k balls constituting W should be surrounded by a fence of a vacant region. The only way this is possible is if the region formed by placing balls of radius $2r$ at each of these k points does not contain any point of the Poisson process other than these k. It is easy to observe that the volume of this region is less when the k points are close together than when the k points are quite separated from each other. Indeed the volume is minimized when all the k points are co-incident with each other. Thus one possible way the event $\{X(W) = k\}$ can occur is when all the k points are clustered very close to each other and around these k points there is an 'annulus' of width approximately $2r$ where there is no Poisson point. In case the density of the Boolean model is large, then having k points very close to each other is quite feasible, although having the annulus free of Poisson points is very unlikely. However, any other situation for the occurrence of $\{X(W) = k\}$ involves a larger volume left free of Poisson points and placing k points in a larger volume and this would have an even lower probability of occurrence than in the situation described previously. Thus in a high-density Boolean model $\{X(W) = k\}$ is a 'rare event', and if it occurs then it is quite likely that it is due to the first scenario of k points being very close to each other and an annulus around these points of width $2r$ being free of Poisson points. In the rest of this section we make these heuristics rigorous.

Here the Poisson process X we consider will be conditioned to have a point at the origin, i.e. $0 \in X$, where 0 is the origin. Also, the radius of the balls comprising the Boolean model will be of a fixed radius r. Since X will always contain a point at the origin, instead of components of the origin W with k Poisson points, for ease of calculation, we shall subsequently consider components with $k + 1$ points.

Before we state and prove the main result, we introduce some notation. One possible situation whereby the first scenario described earlier could occur is to have k points besides the point at the origin in a ball of radius α and outside this ball is an annulus of width $2r$ where there are no Poisson points. The probability of this is clearly

$$
\frac{(\lambda \pi_d \alpha^d)^k}{k!} \exp(-\lambda \pi_d \alpha^d) \exp(-\lambda[\pi_d(\alpha + 2r)^d - \pi_d \alpha^d])
$$

$$
= \frac{(\lambda \pi_d \alpha^d)^k}{k!} \exp(-\lambda \pi_d(\alpha + 2r)^d). \tag{5.6}
$$

An easy calculation shows that as a function of α, the expression on the right side of the equality (5.6) is maximized at $\alpha = \alpha_{k/\lambda}$ where α_c denotes the positive solution of the equation

$$
\alpha(\alpha + 2r)^{d-1} = \frac{c}{\pi_d}. \tag{5.7}
$$

Let

$$
p(\lambda, k) = \sup_{\alpha > 0} \frac{(\lambda \pi_d \alpha^d)^k}{k!} \exp(-\lambda \pi_d(\alpha + 2r)^d).
$$

In case $\alpha_{k/\lambda}$ is so small that the balls of radius r at the k Poisson points constitute a connected set, then clearly

$$P_\lambda\{X(W) = k + 1\} \geq p(\lambda, k).$$

We shall show that $p(\lambda, k)$ is quite close to $P_\lambda\{X(W) = k + 1\}$ for large λ.

In this connection a quantity to study is the relative density of the component W with respect to the ambient density λ of the underlying Poisson process. For any region $C \subseteq \mathbb{R}^d$ containing at least one Poisson point, let diam(C) denote the largest distance between any pair of Poisson points lying in C; i.e. diam$(C) := \sup\{d(x, x') : x, x' \text{ Poisson points in } C\}$. In case C contains only one Poisson point, diam(C) is taken to be 0. Thus, for a bounded component W, a ball centred at any Poisson point of W and of radius diam(W) contains all the Poisson points of the component W and it also contains the convex hull W_H of the Poisson points of W. We define the *density* of the component W as $X(W)/\ell(W_H)$. Thus the ratio $\phi(\lambda) := X(W)/\lambda\ell(W_H)$ defines the *relative density* of the component W with respect to the ambient density λ. We observe that

$$\phi(\lambda) \geq \frac{X(W)}{\lambda\pi_d(\text{diam}(W))^d}. \tag{5.8}$$

Theorem 5.3 *As $\lambda \to \infty$, for a fixed $k \geq 1$, we have*

(i) $\alpha_{k/\lambda} \to 0$,

(ii) $P_\lambda\{X(W) = k + 1\} = \exp\left(-\left[\lambda\pi_d(2r)^d + (d - 1)k \log\frac{\lambda}{k} + O(1)\right]\right)$.
   Moreover, given any $\epsilon > 0$ and $M > 0$, there exist $0 < a < b < \infty$ (depending only on ϵ) and $\lambda' < \infty$ such that, for all $\lambda \geq \lambda'$,

(iii) $P_\lambda\left\{a < \dfrac{\text{diam}(W)}{\alpha_{k/\lambda}} < b \,\middle|\, X(W) = k + 1\right\} \geq 1 - \epsilon,$

(iv) $P_\lambda\{\phi(\lambda) \geq M | X(W) = k + 1\} \geq 1 - \epsilon.$

Statement (iv) says that typically a bounded component in a high-density Boolean model is formed by the Poisson points constituting the component being very close to each other. The density of the component is of a larger order than the ambient density of the process. This is known as the phenomenon of *compression*.

We note, from (5.7), $\alpha_{k/\lambda} \to 0$ as $\lambda \to \infty$ and this proves (i) of the theorem. Also, for any $\lambda > 0$, there exists $\nu_\lambda > 0$ such that $\nu_\lambda \to 0$ as $\lambda \to \infty$ and

$$\frac{k}{\lambda\pi_d(\nu_\lambda + 2r)^{d-1}} \leq \alpha_{k/\lambda} \leq \frac{k}{\lambda\pi_d(2r)^{d-1}}. \tag{5.9}$$

This shows that for large λ, $\alpha_{k/\lambda}$ is approximately $k/\lambda\pi_d(2r)^{d-1}$. For the remainder of the proof we need to do much more work. Since we make extensive use of Stirling's formula, we state the version we use (see Feller 1978, pp. 52–54).

Stirling's formula

$$\lim_{n\to\infty} \frac{n!}{\sqrt{2\pi}n^{n+1/2}e^{-n}} = 1.$$

Moreover, for every $n \geq 1$,

$$\sqrt{2\pi}n^{n+1/2}e^{-n}\exp(1/(12n+1)) \leq n!$$
$$\leq \sqrt{2\pi}n^{n+1/2}e^{-n}\exp(1/(12n)).$$

We begin with an easy lower bound for $P_\lambda\{X(W) = k+1\}$.

Lemma 5.1   *There exists $\lambda_0 < \infty$ such that, for $\lambda \geq \lambda_0$*

$$P_\lambda\{X(W) = k+1\}$$
$$\geq \exp\left(-\left[\lambda\pi_d(2r)^d + (d-1)k\log\frac{\lambda}{k}\right.\right.$$
$$\left.\left. + (d-1)k\log(e\pi_d(2r)^d) + C_1\frac{k^2}{\lambda} + C_2\log k\right]\right),$$

for positive constants C_1 and C_2 independent of λ.

Proof   Let S denote the unit ball centred at the origin and, for $\alpha > 0$, αS denote the ball of radius α centred at the origin. If λ is so large that $k/\lambda\pi_d(2d)^{d-1} < 2r$, then all the Poisson points in the ball $(k/\lambda\pi_d(2d)^{d-1})S$ are in the same component and thus we have

$$P_\lambda\{X(W) = k+1\}$$
$$\geq P_\lambda\left\{X\left(\frac{k}{\lambda\pi_d(2r)^{d-1}}S\right) = k \text{ and}\right.$$
$$\left.X\left(\left(\frac{k}{\lambda\pi_d(2r)^{d-1}} + 2r\right)S \setminus \frac{k}{\lambda\pi_d(2r)^{d-1}}S\right) = 0\right\}$$
$$= \frac{1}{k!}\exp\left(-\lambda\pi_d\left(\frac{k}{\lambda\pi_d(2r)^{d-1}}\right)^d\right)\left[\lambda\pi_d\left(\frac{k}{\lambda\pi_d(2r)^{d-1}}\right)^d\right]^k$$
$$\times \exp\left(-\lambda\pi_d\left[\left(\frac{k}{\lambda\pi_d(2r)^{d-1}} + 2r\right)^d - \left(\frac{k}{\lambda\pi_d(2r)^{d-1}}\right)^d\right]\right).$$

Now using Stirling's formula in this expression, we have for positive constants C_1 and C_2 independent of k or λ,

$$P_\lambda\{X(W) = k+1\}$$

$$\geq \frac{1}{\sqrt{2\pi k}} \left[e\pi_d \left(\frac{1}{\pi_d(2r)^{d-1}} \right)^d \left(\frac{k}{\lambda} \right)^{d-1} \right]^k$$

$$\times \exp \left(-\left[\lambda\pi_d(2r)^d + d\pi_d \left(\frac{1}{\pi_d(2r)^{d-1}} \right) kr^{d-1} + C_1\frac{k^2}{\lambda} \right] \right)$$

$$\geq \exp \left(-\left[\lambda\pi_d(2r)^d + (d-1)k \log\frac{\lambda}{k} \right.\right.$$

$$\left.\left. + (d-1)k \log(e\pi_d(2r)^d) + C_1\frac{k^2}{\lambda} + C_2 \log k \right] \right). \tag{5.10}$$

$\square$

Note that the last three terms in the exponential of the lower bound in Lemma 5.1 contribute $\exp(-[(d-1)k \log(e\pi_d(2r)^d) + C_1(k^2/\lambda) + C_2 \log k])$ which, for fixed $k \geq 1$ and as $\lambda \to \infty$, is of the order of $\exp(-O(1))$. In other words, the bound obtained in (5.10) is the expression we required in (ii) of the theorem. Thus to complete the proof of the theorem we need to show that the $k+1$ Poisson points forming the component W cannot be separated any more than required to compute (5.10). We prove this in a sequence of lemmas. The first lemma considers the case when the component W is bounded and has a large diameter.

Lemma 5.2 *For every $\beta < \infty$, there exist $l < \infty$ and $\lambda_1 < \infty$ such that for $\lambda > \lambda_1$ we have*

$$P_\lambda\{l \leq d(W) < \infty\} \leq \exp(-\lambda\beta).$$

Proof The proof proceeds by a lattice approximation. Suppose $l \leq d(W) < \infty$. Consider a lattice $\mathbb{L}$ of width a with $0 < a < r/4$. Since $d(W) \geq l$, there must exist a connected component $\delta_0(W)$ (say) of the boundary $\delta(W)$ such that $\delta_0(W)$ encloses the origin and its diameter $d(\delta_0(W))$ is at least l. Let $\delta_c(W)$ be the collection of cells of the lattice $\mathbb{L}$ which intersects $\delta_0(W)$. Since $d(\delta_0(W)) \geq l$, the number of cells in $\delta_c(W)$ is at least l/a and also $\cup_{C \in \delta_c(W)} C$ is a connected set. Moreover, by our choice of the lattice width a, no Poisson point can lie on any cell in $\delta_c(W)$.

Now for every $n \geq 1$ consider the set $\mathcal{K}_n$ of all collections $\mathcal{C}$ of cells of the lattice $\mathbb{L}$ such that (i) the number of cells in $\mathcal{C}$ is n, (ii) $\cup_{C \in \mathcal{C}} C$ is a connected set, and (iii) $\cup_{C \in C} C$ contains a $(d - 1)$-dimensional surface which encloses the origin. (In other words, the set $\mathcal{K}_n$ contains all collections C of cells which have cardinality n and which arise as a collection $\delta_c(W)$ of a bounded component W.) A counting argument as in Theorem 1.1 yields that the cardinality κ_n of $\mathcal{K}_n$ is at most c^n for some constant $c > 1$ which depends on the dimension d. Thus,

$$P_\lambda\{l \le d(W) < \infty\} \le \sum_{n \ge l/a} \sum_{C \in \mathcal{K}_n} P_\lambda(X(\cup_{C \in C} C) = 0)$$

$$\le \sum_{n \ge l/a} c^n \exp(-\lambda n a^d). \tag{5.11}$$

Now given $\beta < \infty$ choose l such that $\beta = l a^{d-1} \log(2b^{l/a})$; then, for λ such that $1 - c \exp(-\lambda a^d) > 1/2$, we have $\sum_{n \ge l/a} c^n \exp(-\lambda n a^d) < \exp(-\lambda \beta)$. This along with (5.11) proves the lemma. $\square$

REMARK: Under the conditions of Lemma 5.2 we also obtain

$$P_\lambda\{l \le \text{diam}(W) < \infty\} \le \exp(-\lambda \beta). \tag{5.12}$$

Now we turn our attention to components of smaller diameter. The next lemma will be repeatedly applied in the subsequent lemmas.

Lemma 5.3 *Let $\mu > 1$ and define, for $y > 0$,*

$$\psi_\mu(y) = \frac{\pi_{d-1}(2r)^{d-1} y}{4} - \log(e\pi_d \mu^d y^d).$$

There exists a constant $C_3 = C_3(\mu, r)$ such that for $y(k/\lambda) < C_3$, we have

$$P_\lambda\left\{X(W) = k + 1, \ y\frac{k}{\lambda} < \text{diam}(W) \le \mu y\frac{k}{\lambda}\right\}$$

$$\le \exp\left(-\left[\lambda\pi_d(2r)^d + (d-1)k \log\frac{\lambda}{k} + k\psi_\mu(y)\right]\right). \tag{5.13}$$

Proof Let W_p denote the set of all Poisson points of X (including the origin) in the ball $\mu y(k/\lambda)S$. Define

$$A := \left\{X\left(\mu y\frac{k}{\lambda}S\right) = k\right\}$$

$$B := \left\{y\frac{k}{\lambda} < \text{diam}(W_p) \le \mu y\frac{k}{\lambda}\right\}.$$

If $\mu y(k/\lambda) < 2r$, we have

$$P_\lambda \left\{ X(W) = k+1, \ y\frac{k}{\lambda} < \text{diam}(W) \leq \mu y\frac{k}{\lambda} \right\}$$

$$= E_\lambda \left(P_\lambda \left\{ X(W) = k+1, \ y\frac{k}{\lambda} < \text{diam}(W) \leq \mu y\frac{k}{\lambda} | W_p \right\} \right)$$

$$= E_\lambda \left(P_\lambda \left\{ X\left(\cup_{w \in W_p} S(w, 2r) \setminus \mu y\frac{k}{\lambda} S \right) = 0 | W_p \right\} 1_A 1_B \right), \qquad (5.14)$$

where as usual $S(w, 2r)$ denotes the ball of radius $2r$ centred at w.

To estimate the last term in (5.14), we need to estimate the Lebesgue measure of $\cup_{w \in W_p} S(w, 2r)$. Let w_1 and w_2 be two points of W_p such that w_1 and w_2 are farthest apart among all pairs of points of W_p; i.e., $d(w_1, w_2) = \text{diam}(W_p)$. Let H_1 and H_2 be hyperplanes through w_1 and w_2 respectively, both of which are perpendicular to the line L passing through w_1 and w_2. Clearly all points of W_p lie in the slab T between the two hyperplanes H_1 and H_2. Let w_0 be the point of W_p which lies farthest (in terms of the perpendicular distance) from the line L, and let H_0 be the hyperplane passing through w_0 and which is perpendicular to the shortest line joining w_0 to L. Clearly each of the hyperplanes H_1, H_2 and H_0 divides the space $\mathbb{R}^d$ into a pair of distinct half-spaces such that one of each pair H_1^+, H_2^+ and H_0^+ (say) contains no point of W_p. The Lebesgue measure of $\cup_{w \in W_p} S(w, 2r)$ is clearly larger than the sum of the Lebesgue measures of

(i) a semisphere centred at w_1 of radius $2r$ and lying completely in H_1^+ (region I in Figure 5.2)

(ii) a semisphere centred at w_2 of radius $2r$ and lying completely in H_2^+ (region II in Figure 5.2) and

(iii) the region in the slab T enclosed by the semisphere centred at w_0 of radius $2r$ and lying completely in H_0^+ (region III in Figure 5.2).

Thus,

$$\ell(\cup_{w \in W_p} S(w, 2r))$$

$$\geq \ell(H_1^+ \cap S(w_1, 2r)) + \ell(H_2^+ \cap S(w_2, 2r)) + \ell(T \cap H_0^+ \cap S(w_0, 2r))$$

$$= \pi_d (2r)^d + \ell(T \cap H_0^+ \cap S(w_0, 2r)). \qquad (5.15)$$

To estimate $\ell(T \cap H_0^+ \cap S(w_0, 2r))$, we first give an argument in two dimensions. A Pythagorean calculation gives that in two dimensions, the disc $S(w_0, 2r)$ centred at w_0 makes an intercept of length at least $\sqrt{(2r)^2 - (y\frac{k}{\lambda})^2}$

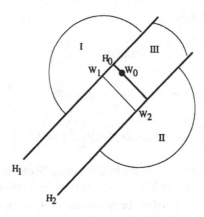

Figure 5.2. Regions I, II and III.

with each of the two lines H_1 and H_2. Thus, if $\mathrm{diam}(W_p) \geq y\frac{k}{\lambda}$, then $T \cap H_0^+ \cap S(w_0, 2r)$ contains a rectangle of dimension $\sqrt{(2r)^2 - (y\frac{k}{\lambda})^2} \times y(k/\lambda)$. Hence, for λ large enough $\ell(T \cap H_0^+ \cap S(w_0, 2r)) \geq 2ry(k/4\lambda)$.

In general, for higher dimensions, the intercepts on each of the hyperplanes H_1 and H_2 will be $(d - 1)$-dimensional semispheres of $(d - 1)$-dimensional Lebesgue measure at least $\pi_d\big(\sqrt{(2r)^2 - (y\frac{k}{\lambda})^2}\big)^{d-1}$. Thus, if $\mathrm{diam}(W_p) \geq y(k/\lambda)$, then $T \cap H_0^+ \cap S(w_0, 2r)$ contains a d-dimensional semi-cylinder whose base is a $(d - 1)$-dimensional semisphere and whose height is $y\frac{k}{\lambda}$. Thus, we have, for large λ,

$$\ell(T \cap H_0^+ \cap S(w_0, 2r)) \geq \pi_{d-1}(2r)^{d-1}y\frac{k}{4\lambda}. \qquad (5.16)$$

We now use independence obtained from the fact that the event A depends only on the configuration inside the ball $\mu y\frac{k}{\lambda}S$, whereas the event $\{X(\cup_{w \in W_p}S(w, 2r)\backslash\mu y(k/\lambda)S) = 0\}$ depends only on the configuration outside the ball $\mu yk/\lambda S$, to have from (5.15) and (5.16),

$$E_\lambda\left(P_\lambda\left\{X\left(\cup_{w \in W_p}S(w, 2r)\backslash\mu y\frac{k}{\lambda}S\right) = 0|W_p\right\}1_A1_B\right)$$

$$\leq E_\lambda\left(P_\lambda\left\{X\left(\cup_{w \in W_p}S(w, 2r)\backslash\mu y\frac{k}{\lambda}S\right) = 0|W_p\right\}1_A\right)$$

$$\leq \exp\left(-\lambda\left[\pi_d(2r)^d + \pi_{d-1}(2r)^{d-1}y\frac{k}{4\lambda} - \ell\left(\mu y\frac{k}{\lambda}S\right)\right]\right)P_\lambda(A)$$

$$= \exp\left(-\lambda\left[\pi_d(2r)^d + \pi_{d-1}(2r)^{d-1}y\frac{k}{4\lambda} - \pi_d\left(\mu y\frac{k}{\lambda}\right)^d\right]\right)$$

$$\times \frac{[\lambda\pi_d(\mu y\frac{k}{\lambda})^d]^k}{k!} \exp\left(-\lambda\pi_d\left(\mu y\frac{k}{\lambda}\right)^d\right)$$

$$\leq \exp\left(-\left[\lambda\pi_d(2r)^d + (d-1)\log\frac{\lambda}{k} + k\psi_\mu(y)\right]\right), \tag{5.17}$$

where we have used Stirling's formula in the last inequality. From (5.14) and (5.17), we see that the lemma holds for any C_3 with $\mu C_3 < 2r$. $\qquad\square$

We now use the counting method of Lemma 5.2 and the geometric argument of Lemma 5.3 to study the case when the component is of 'medium' size.

Lemma 5.4 *For every $0 < \delta$ and $l < \infty$, there exist constants $C_4 > 0$ and $\lambda_2 < \infty$ such that for $\lambda \geq \lambda_2$*

$$P_\lambda\{\delta \leq \text{diam}(W) \leq l\} \leq \exp(-\lambda[\pi_d(2r)^d + C_4]).$$

Proof As in the geometric argument given in Lemma 5.3, here, if $\text{diam}(W) \geq \delta$, then we require that two semispheres of radius $2r$ each separated by a distance of at least δ and a region formed by the intersection of a semisphere and a slab of thickness at least δ have to be free of Poisson points. Let this region be denoted by A. Clearly $A \subseteq [-l - 2r, l + 2r]^d$. The argument given to justify (5.16) yields in this case that the Lebesgue measure of the region A satisfies $\ell(A) \geq \pi_d(2r)^d + c_1\pi_{d-1}(2r)^{d-1}\delta$, for some constant $0 < c_1 < 1$. (Note that here δ plays the role of $y(k/\lambda)$ of the previous lemma.) Introducing a lattice approximation as in Lemma 5.2 with a lattice $\mathbb{L}$ of width $a > 0$, we have the inner lattice approximation A_a of A given by $A_a := \cup\{C : C$ a cell of the lattice $\mathbb{L}$ and $C \subseteq A\}$. For $a < c_2$ where $c_2 > 0$ is a sufficiently small constant, we obtain a constant $c_3 > 0$ depending on c_1 and c_2 such that $\ell(A_a) \geq \pi_d(2r)^d + c_3\delta$. Now $A_a \subseteq [-l - 2r, l + 2r]^d$; thus for a fixed a, there are only finitely many possible choices of A_a. Hence a summation over all possible A_a's yields

$$P_\lambda\{\delta \leq \text{diam}(W) \leq l\} \leq \sum_{A_a} P_\lambda\{A_a \text{ contains no Poisson point}\}, \tag{5.18}$$

where the summation is over all A_a's such that A_a's are unions of cells of the lattice $\mathbb{L}$ and $\ell(A_a) \geq \pi_d(2r)^d + c_3\delta$. Let $N_a(= N_a(l, \delta, r))$ be the total number of such choices of A_a possible. From (5.18) we have

$$P_\lambda\{\delta \le \text{diam}(W) \le l\} \le \sum_{A_a} \exp(-\lambda[\pi_d(2r)^d + c_3\delta])$$

$$\le N_a \exp(-\lambda[\pi_d(2r)^d + c_3\delta]). \qquad (5.19)$$

Taking $C_4 = c_3\delta/2$, we obtain $\lambda_2 < \infty$ such that for $\lambda \ge \lambda_2$,

$$\exp(-\lambda c_3\delta + \log N_a(l, \delta, r)) \le \exp(-\lambda C_4). \qquad (5.20)$$

Combining (5.19) and (5.20) yields the lemma.   □

In the next lemma we apply Lemma 5.3 to take care of small components.

Lemma 5.5   *Given $\beta < \infty$, there exist $C_5 > 0$, $\delta > 0$ (depending only on β) and λ_3 such that, for $\lambda \ge \lambda_3$, we have*

$$P_\lambda\left\{X(W) = k+1, C_5\frac{k}{\lambda} \le \text{diam}(W) \le \delta\right\}$$

$$\le \exp\left(-\left[\lambda\pi_d(2r)^d + (d-1)k\log\frac{\lambda}{k} + \beta k\right]\right).$$

Proof   First we fix $\mu > 1$ and obtain the constant C_3 as in Lemma 5.3. Now we fix $\delta < C_3$. Observe that the function ψ_μ of Lemma 5.3 satisfies $\psi_\mu(y) \to \infty$ as $y \to \infty$, so that we may choose a constant C_5 such that the following two conditions hold:

$$\psi_\mu(C_5) > \beta + \log 2, \qquad (5.21)$$

$$\sum_{j=1}^{\infty} \exp(-k[\psi_\mu(C_5\mu^j) - \psi_\mu(C_5)]) \le 1. \qquad (5.22)$$

Let $\lambda_3 < \infty$ be such that $C_5(k/\lambda) < \delta$ and for some $j \ge 1$, $\delta < C_5\mu^j(k/\lambda) < C_3$ whenever $\lambda \ge \lambda_3$. We now partition the interval $[C_5(k/\lambda), \delta]$ by intervals of the type $[C_5\mu^j(k/\lambda), C_5\mu^{j+1}(k/\lambda)]$, $j = 0, 1, \dots$. Let N be the smallest integer such that $[C_5(k/\lambda), \delta] \subseteq \cup_{j=0}^{N}[C_5\mu^j(k/\lambda), C_5\mu^{j+1}(k/\lambda)]$. Since $C_5\mu^j(k/\lambda) \to \infty$ as $j \to \infty$ for fixed λ, we have that N is finite. Now observe that

$$P_\lambda\left\{X(W) = k+1, C_5\frac{k}{\lambda} \le \text{diam}(W) \le \delta\right\}$$

$$\le \sum_{j=0}^{N} P_\lambda\left\{X(W) = k+1, C_5\mu^j\frac{k}{\lambda} \le \text{diam}(W) \le C_5\mu^{j+1}\frac{k}{\lambda}\right\}. \qquad (5.23)$$

Applying Lemma 5.3 to each of the terms inside the summation of (5.23) we have, for $\lambda \geq \lambda_3$,

$$P_\lambda \left\{ X(W) = k+1, \ C_5 \frac{k}{\lambda} \leq \text{diam}(W) \leq \delta \right\}$$

$$\leq \sum_{j=0}^{N} \exp\left(-\left[\lambda \pi_d (2r)^d + (d-1)k \log \frac{\lambda}{k} + k\psi_\mu(C_5\mu^j)\right]\right)$$

$$\leq \exp\left(-\left[\lambda \pi_d (2r)^d + (d-1)k \log \frac{\lambda}{k} + k\psi_\mu(C_5)\right]\right)$$

$$+ \sum_{j=1}^{N} \exp\left(-\left[\lambda \pi_d (2r)^d + (d-1)k \log \frac{\lambda}{k} + k\psi_\mu(C_5\mu^j)\right]\right)$$

$$\leq 2\exp\left(-\left[\lambda \pi_d (2r)^d + (d-1)k \log \frac{\lambda}{k} + k\psi_\mu(C_5)\right]\right)$$

$$\leq \exp\left(-\left[\lambda \pi_d (2r)^d + (d-1)k \log \frac{\lambda}{k} + \beta k\right]\right), \qquad (5.24)$$

where the last two inequalities follow from (5.22) and (5.21). This completes the proof of the lemma. □

In the last four lemmas, we have covered the case when $\text{diam}(W) \geq C_5(k/\lambda)$. The next lemma considers the case when the component is very small.

Lemma 5.6 *Given any $\beta < \infty$, there exist $C_6 > 0$ and $\lambda_4 < \infty$ such that, for $\lambda \geq \lambda_4$, we have $C_6(k/\lambda) < \delta$ and*

$$P_\lambda \left\{ X(W) = k+1, \ \text{diam}(W) \leq C_6\frac{k}{\lambda} \right\}$$

$$\leq \exp\left(-\left[\lambda \pi_d (2r)^d + (d-1)k \log \frac{\lambda}{k} + \beta k\right]\right).$$

Proof The proof of this lemma is similar to that of the previous lemma and as such we just sketch the proof. Here we observe that $\psi_\mu(y) \to \infty$ as $y \to 0$. Thus we have to choose $C_6 < C_5$ sufficiently small such that

$$\psi_\mu(C_6) > \beta + \log 2, \qquad (5.25)$$

$$\sum_{j=1}^{\infty} \exp(-k[\psi_\mu(C_6\mu^{-j}) - \psi_\mu(C_6)]) \leq 1, \qquad (5.26)$$

where $\mu > 1$ is a previously fixed quantity. Also, we may choose λ_4 large such that $C_6(k/\lambda) < C_3$ for all $\lambda \geq \lambda_4$ and that ensures that we may apply Lemma 5.3. Covering the interval $(0, C_6(k/\lambda)]$ by intervals of the type $[C_6\mu^{-(j+1)}(k/\lambda), C_6\mu^{-j}(k/\lambda)]$, $j = 0, 1, \ldots$, we obtain an inequality as in (5.23)

$$
P_\lambda \left\{ X(W) = k + 1, \ \text{diam}(W) \leq C_6 \frac{k}{\lambda} \right\}
$$

$$
\leq \sum_{j=0}^{\infty} P_\lambda \left\{ X(W) = k + 1, \ C_6\mu^{-(j+1)} \frac{k}{\lambda} \leq \text{diam}(W) \leq C_5\mu^{-j} \frac{k}{\lambda} \right\}.
$$

A calculation similar to (5.24) using (5.26) and (5.25) completes the proof of the lemma.   □

Proof of Theorem 5.3   First observe that taking $\mu_0 = C_5/C_6$, where C_5 and C_6 are as in Lemmas 5.5 and 5.6, we have $\mu_0 > 1$. Also, from Lemma 5.3, for λ_5 large enough such that $C_5(k/\lambda) < C_3(\mu_0, r)$ for all $\lambda \geq \lambda_5$, we have

$$
P_\lambda \left\{ X(W) = k + 1, \ C_6 \frac{k}{\lambda} \leq \text{diam}(W) \leq C_5 \frac{k}{\lambda} \right\}
$$

$$
\leq \exp\left(-\left[\lambda \pi_d (2r)^d + (d-1)k \log \frac{\lambda}{k} + k\psi_{\mu_0}(C_6) \right] \right). \quad (5.27)
$$

Now let us collect all our observations. Fix β such that

$$
\beta > \pi_d (2r)^d \quad \text{and} \quad 1 + \exp(-\beta k) \leq \exp\left(\frac{\beta k}{6} \right) \quad (5.28)
$$

(the need for this choice will be apparent later), and accordingly obtain $0 < l(= l(\beta))$ as in Lemma 5.2, $0 < \delta(= \delta(\beta))$ and $C_5(= C_5(\beta)) < \infty$ as in Lemma 5.5 and $0 < C_6(= C_6(\beta)) < C_5$ as in Lemma 5.6. For $X(W) = k + 1$, we see that, if $\lambda \geq \max\{\lambda_1, \lambda_2, \lambda_3, \lambda_4, \lambda_5\}$,

(i) (5.12) accounts for the case when $\text{diam}(W) \geq l$ for some l,
(ii) Lemma 5.4 accounts for the case when $\delta \leq \text{diam}(W) \leq l$,
(iii) Lemma 5.5 accounts for the case when $C_5(k/\lambda) \leq \text{diam}(W) \leq \delta$,
(iv) Lemma 5.6 accounts for the case when $\text{diam}(W) \leq C_6(k/\lambda)$,
(v) (5.27) accounts for the case when $C_6(k/\lambda) \leq \text{diam}(W) \leq C_5(k/\lambda)$.

Thus we have accounted for the entire possible range of $\text{diam}(W)$. Moreover, the upper bound on the probability obtained in each of these cases is at most $\exp(-[\lambda \pi_d (2r)^d + (d-1)k \log(\lambda/k) + O(1)])$. Thus, we have obtained

$$
P_\lambda \{ X(W) = k + 1 \} \leq \exp\left(-\left[\lambda \pi_d (2r)^d + (d-1)k \log \frac{\lambda}{k} + O(1) \right] \right)
$$

as $\lambda \to \infty$. This along with the lower bound obtained in Lemma 5.1 proves (ii) of the theorem.

The choice of β comes in to prove (iii). From Lemma 5.2, Lemma 5.4 and Lemma 5.5, for all sufficiently large λ,

$$P_\lambda \left\{ X(W) = k + 1, \ \text{diam}(W) \geq C_5 \frac{k}{\lambda} \right\}$$
$$\leq \exp \left(- \left[\lambda \pi_d (2r)^d + (d-1)k \log \frac{\lambda}{k} + \frac{\beta k}{2} \right] \right) \qquad (5.29)$$

and from Lemma 5.6

$$P_\lambda \left\{ X(W) = k + 1, \ \text{diam}(W) \leq C_6 \frac{k}{\lambda} \right\}$$
$$\leq \exp \left(- \left[\lambda \pi_d (2r)^d + (d-1)k \log \frac{\lambda}{k} + \beta k \right] \right). \qquad (5.30)$$

Combining the lower bound obtained in Lemma 5.1 with (5.29) and (5.30), and using the second part of (5.28) in the choice of β, we obtain for all sufficiently large λ

$$\frac{P_\lambda \{ X(W) = k + 1, \ \text{diam}(W) \leq C_6(k/\lambda) \text{ or } \text{diam}(W) \geq C_5(k/\lambda) \}}{P_\lambda \{ X(W) = k + 1 \}}$$
$$\leq \exp \left(- \left[\lambda \pi_d (2r)^d + (d-1)k \log \frac{\lambda}{k} + \frac{\beta k}{3} \right] \right)$$
$$\times \exp \left(\left[\lambda \pi_d (2r)^d + (d-1)k \log \frac{\lambda}{k} \right. \right.$$
$$+ (d-1)k \log(e\pi_d (2r)^d) + C_1 \frac{k^2}{\lambda} + C_2 \log k \left. \left. \right] \right)$$
$$\leq \exp \left(- \left[\frac{\beta k}{3} - (d-1)k \log(e\pi_d (2r)^d) \right. \right.$$
$$+ C_1 \frac{k^2}{\lambda} + C_2 \log k \left. \left. \right] \right). \qquad (5.31)$$

Given $\epsilon > 0$, we choose β such that (5.28) is satisfied and

$$\exp \left(- \left[\frac{\beta k}{3} - (d-1)k \log(e\pi_d (2r)^d) + C_1 \frac{k^2}{\lambda} + C_2 \log k \right] \right) \leq \epsilon \quad (5.32)$$

for all λ sufficiently large. Without loss of generality suppose that this λ is so large that $k/(2\lambda\pi_d (2r)^{d-1}) \leq k/(\lambda\pi_d (v_\lambda + 2r)^{d-1})$, where v_λ is as in (5.9). Now choose

$$a = C_6 \pi_d (2r)^{d-1} \quad \text{and} \quad b = 2C_5 \pi_d (2r)^{d-1}, \qquad (5.33)$$

according to this choice of β. For β satisfying (5.28) and (5.32) and a and b as in (5.33), from (5.30) we have

$$P_\lambda \left\{ a < \frac{\text{diam}(W)}{\alpha_{k/\lambda}} < b | X(W) = k+1 \right\} \geq 1 - \epsilon,$$

and this proves (iii).

To show the compression phenomenon we first observe that (5.9) and (iii) yield that, for all sufficiently large λ, with P_λ probability at least $1 - \epsilon$ we have

$$\frac{k}{\lambda \pi_d \, \text{diam}(W)^d} \geq \frac{k}{b \lambda \pi_d \left(\frac{k}{\lambda \pi_d (2r)^{d-1}} \right)^d}.$$

Given $\epsilon > 0$ and $M > 0$, we may choose λ so large that the term on the right side of the above expression is larger than M and this, along with (5.8), proves (iv) of the theorem. $\square$

5.3 Rarefaction

In the previous section the compression phenomenon was obtained by considering balls of a fixed radius r. A natural question is what happens when we have varying radius, and in this section we investigate this.

We shall assume throughout this section that the radius random variable ρ is non-degenerate and takes exactly two different values; i.e. for $0 < r < R < \infty$, we have

$$P(\rho = R) = 1 - P(\rho = r) = p, \tag{5.34}$$

for some $0 < p < 1$. Throughout this section, big balls will always refer to balls of radius R and small balls will always refer to balls of radius r.

Before we present the formal details, we give some intuitive ideas about the structure of a bounded component in the Boolean model (X, ρ, λ) when the density λ is very large and ρ satisfies (5.34). The Boolean model (X, ρ, λ) is assumed to include a Poisson point at the origin.

In case the Poisson point at the origin accommodates a big ball and the component W consists of k big balls (besides the ball centred at the origin) and l small balls, then a possible structure of the component W is that the centres of the k big balls and the big ball at the origin are all clustered near the origin, while the small balls are distributed 'uniformly' in the region formed by the big balls such that none of the small balls protrude outside the region formed by the big balls. In this case an annulus of width $2r$ around the region formed by the big balls has to be devoid of Poisson points which are centres of small balls, and an annulus of width $2R$ around the region formed by the big balls has to be devoid of Poisson points which are centres of big balls. A little thought shows that it is this structure which would minimize the volume of the annular region which needs to be free of Poisson points.

In the argument above, we assumed that the origin is the centre of a big ball. However, if the origin is the centre of a small ball and component W contains exactly k big balls and l small balls (besides the ball centred at the origin), then, provided $k \geq 1$, the structure of the component W will not be vastly different from that described in the previous paragraph. Indeed what could happen is that the centres of the big balls are all clustered together (which need not be around the origin) and the $l + 1$ small balls (including the small ball at the origin) are uniformly distributed in the region formed by the big balls such that none of the small balls protrude outside this region. Thus the difference between the structure of the cluster obtained in this case and that obtained in the previous paragraph involves just a change in the position of the origin.

In case the component W consists of $k + 1$ big balls and there is no small ball centred in W, then as in the previous section, we will have a compression phenomenon with all the Poisson points (besides the origin) compressed in a small region around the origin. A similar phenomenon will be observed when the component W consists of $l + 1$ small balls and no big ball is centred in W.

Returning to the situation when the component W contains both big and small balls and a big ball is centred at the origin, we see that the l small balls are distributed in the region formed by the $k + 1$ big balls. This region formed by the $k + 1$ big balls contains a spherical region inside it of Lebesgue measure at least $\pi_d R^d$. The centres of the small balls may be placed anywhere inside this region at a distance at least r from the boundary of the region to guarantee that the small balls do not protrude out of this region. This means that the small balls may be centred in a region whose Lebesgue measure is at least $\pi_d (R - r)^d$. Thus looking at the whole picture we see that the $k + l + 1$ Poisson points forming the component W are distributed in a region of Lebesgue measure at least $\pi_d (R - r)^d$ with $k + 1$ of these points compressed together while the remaining l are distributed uniformly in this region. This would yield a *rarefaction* phenomenon as $\lambda \to \infty$, because a region of Lebesgue measure at least $\pi_d (R - r)^d$ should typically accommodate $\lambda \pi_d (R - r)^d$ Poisson points.

In this section we will state and prove the result only for the case when the origin is the centre of a big ball and there is at least one small ball present in the component W. The case when the component contains at least one big ball and the origin is the centre of a small ball should follow after tedious technical details from the previous case and it involves conditioning on the position of a big ball. We shall omit this case. The details of the case when the component

W consists only of big balls or only of small balls are similar to those described in the previous section and as such we omit this case too.

Before we introduce the relevant notation, it may be observed that the intuitive reasoning given above needs the existence of at least two different sized balls. Thus the reasoning is valid for any general non-degenerate bounded random variable ρ. However, the technical details of the result extend only to non-degenerate ρ which has support in $[r, R]$ for some $0 < r < R < \infty$. We do not venture to prove the result for ρ other than that satisfying (5.34).

Let $(Y, R, \lambda p)$ be a Poisson Boolean model conditioned to have a point at the origin and $(Z, r, \lambda(1 - p))$ be another Poisson Boolean model, where r, R and p are as in (5.34) and $(Y, R, \lambda p)$ and $(Z, r, \lambda(1 - p))$ are independent processes. The superposition of these two Boolean models is a Boolean model which is equivalent in law to a Poisson Boolean model of density λ and radius random variable ρ conditioned to have a point at the origin with a ball of radius R. Throughout this section $0, y_1, y_2, \ldots$ represent points of the point process Y and $z_1, z_2, \ldots$ represent points of the point process Z. Let W_Y and W_Z denote the occupied components of the origin in $(Y, R, \lambda p)$ and $(Z, r, \lambda(1-p))$, respectively, and let W denote the occupied component of the origin in the superposition of these models. Clearly $W \supseteq W_Y \cup W_Z$.

For $l \geq 1, k \geq 0$, let

$$\{\#W = (k, l)\} := \{\text{the origin is the centre of a big ball and } W \text{ contains}$$
$$\text{exactly } k + 1 \text{ points of } Y \text{ and } l \text{ points of } Z\}.$$

As a measure of the size, for any set S containing the origin, let

$$\text{rad}(S) = \sup\{d(0, x) : x \text{ is a point of } Y * Z \text{ in } S\},$$
$$\text{rad}_Y(S) = \sup\{d(0, y) : y \text{ is a point of } Y \text{ in } S\},$$
$$\text{rad}_Z(S) = \sup\{d(0, z) : z \text{ is a point of } Z \text{ in } S\},$$

where $d(\cdot, \cdot)$ denotes the Euclidean distance on $\mathbb{R}^d$.

Theorem 5.4 *Let (X, ρ, λ) be a Boolean model with ρ as in (5.34). For $l \geq 0$ and $k \geq 1$ fixed, all functions $a(\lambda)$ with $a(\lambda) \to 0$ as $\lambda \to \infty$ and for every $\epsilon > 0$, we have, as $\lambda \to \infty$,*

(i) $P_\lambda(\#W = (k, l)|\text{the origin is the centre of a big ball})$
   $= \exp(-\lambda \pi_d E(\rho + R)^d + (l - (d - 1)k) \log \lambda + O(1))$
(ii) $P_\lambda(rad(W) > a(\lambda)|\#W = (k, l),$
   the origin is the centre of a big ball$) \to 1,$

(iii) $P_\lambda(\phi(\lambda) < \epsilon | \#W = (k, l),$

the origin is the centre of a big ball) $\to 1,$

where $\phi(\lambda)$ is the relative density as introduced in the previous section.

The proof proceeds as in the last section. In the first lemma we obtain a lower bound, while in the subsequent lemmas we obtain an upper bound. In all the lemmas, we shall be working on the superposed model $(Y, R, \lambda p) * (Z, r, \lambda(1 - p))$ and as such P_λ will incorporate the condition that there is a big ball at the origin. As in the previous section, S denotes the unit ball centred at the origin.

Lemma 5.7 *For k and l fixed, as $\lambda \to \infty$,*

$$P_\lambda\{\#W = (k, l)\}$$
$$\geq \exp(-\lambda \pi_d E(\rho + R)^d + (l - (d - 1)k) \log \lambda + O(1)).$$

Proof Given that there is a big ball at the origin, if the remaining k big balls are placed in the region $\alpha(k/\lambda)S$ where $\alpha = (p\pi_d(2R)^{d-1})^{-1}$, and λ is so large that $\alpha(k/\lambda) < 2R$, then the k big balls, together with the ball at the origin, are in the same component. (Note, as obtained in the previous section, $\alpha(k/\lambda)$ corresponds to the optimal diameter of the Poisson points in the component W when the Boolean model consists only of big balls; i.e. $p = 1$.) Now centre the l small balls in the region $(R - r)S$. If an annular region of width R of the component W_Y formed by the $k + 1$ big balls is free of Poisson points of Y and an annular region of width r is free of Poisson points of Z, then $W = W_Y$ and $\#W = (k, l)$. Thus

$$P_\lambda\{\#W = (k, l)\}$$

$$\geq P_\lambda \left\{ Y\left(\alpha \frac{k}{\lambda} S\right) = k, \; Z((R - r)S) = l, \right.$$

$$Y\left(\left(2R + \alpha\frac{k}{\lambda}\right) S \setminus \alpha\frac{k}{\lambda} S\right) = 0,$$

$$\left. Z\left(\left((R + r) + \alpha\frac{k}{\lambda}\right) S \setminus (R - r)S\right) = 0 \right\}$$

$$= \exp\left(-\lambda \left[p\pi_d \left(\alpha\frac{k}{\lambda}\right)^d + (1 - p)\pi_d(R - r)^d \right]\right)$$

$$\times \frac{[\lambda p\pi_d(\alpha(k/\lambda))^d]^k}{k!} \frac{[\lambda(1 - p)\pi_d(R - r)^d]^l}{l!}$$

$$\times \exp\left(-\lambda p \pi_d \left[\left(2R + \alpha\frac{k}{\lambda}\right)^d - \left(\alpha\frac{k}{\lambda}\right)^d\right]\right)$$

$$\times \exp\left(-\lambda(1-p)\pi_d \left[\left(R + r + \alpha\frac{k}{\lambda}\right)^d - (R-r)^d\right]\right)$$

$$= \exp(-\lambda\pi_d[p(2R)^d + (1-p)(R+r)^d])$$

$$\times \exp\left(-\lambda\pi_d p \sum_{j=1}^{d} \binom{d}{j}\left(\alpha\frac{k}{\lambda}\right)^j (2R)^{d-j}\right)$$

$$\times \exp\left(-\lambda\pi_d(1-p) \sum_{j=1}^{d} \binom{d}{j}\left(\alpha\frac{k}{\lambda}\right)^j (R+r)^{d-j}\right)$$

$$\times \frac{[\lambda p \pi_d(\alpha(k/\lambda))^d]^k}{k!} \frac{[\lambda(1-p)\pi_d(R-r)^d]^l}{l!}$$

$$= \exp(-\lambda\pi_d E(\rho + R)^d)$$

$$\times \exp\left(-\lambda\pi_d \sum_{j=1}^{d} \binom{d}{j}\left(\alpha\frac{k}{\lambda}\right)^j E(\rho + R)^{d-j}\right)$$

$$\times \frac{[\lambda p \pi_d(\alpha(k/\lambda))^d]^k}{k!} \frac{[\lambda(1-p)\pi_d(R-r)^d]^l}{l!}.$$

Using Stirling's formula for $k!$ and $l!$, and noting that the quantity

$$\lambda\pi_d \sum_{j=1}^{d} \binom{d}{j}\left(\alpha\frac{k}{\lambda}\right)^j E(\rho + R)^{d-j}$$

is $O(1)$ as $\lambda \to \infty$, we have

$$P_\lambda\{\#W = (k,l)\}$$

$$\geq \exp\left(-\lambda\pi_d E(\rho + R)^d + (d-1)k\log\frac{k}{\lambda}\right.$$

$$\left. + l\log\frac{\lambda(1-p)}{l} + l\log(e\pi_d(R-r)^d) + O(1)\right)$$

$$\geq \exp\left(-\lambda\pi_d E(\rho + R)^d + (l - (d-1)k)\log\frac{k}{\lambda} + O(1)\right). \qquad \square$$

As in the previous section, we shall show an upper bound on the probability of obtaining $\#W = (k,l)$. In the next two lemmas we show that the probability that there is a Poisson point of either Y or Z at a distance at least R away from the origin is significantly smaller than the lower bound obtained in Lemma 5.7.

Lemma 5.8 *For some positive constants C_1, C_2, C_3 and C_4, we have*

$$P_\lambda \{\#W = (k, l),\ \mathrm{rad}_Y(W) > 2R\}$$

$$\leq \exp\left(-\lambda \pi_d E(\rho + R)^d - \frac{\lambda}{2} p \pi_d \left(\frac{R}{2}\right)^d + C_1 k + C_2 l\right) \quad (5.35)$$

and

$$P_\lambda \{\#W = (k, l),\ \mathrm{rad}_Z(W) > R + r\}$$

$$\leq \exp\left(-\lambda \pi_d E(\rho + R)^d - \frac{\lambda}{2}(1 - p)\pi_d \left(\frac{r}{2}\right)^d + C_3 k + C_4 l\right). \quad (5.36)$$

Proof Let

$$A_i := \{Y((2R)S) = i\} \quad \text{for } i = 0, 1 \ldots, k,$$
$$B_j := \{Z((R + r)S) = j\} \quad \text{for } j = 0, 1, \ldots, l.$$

Since the origin is the centre of a big ball, if $\#W = (k, l)$ and $\mathrm{rad}_Y(W) > 2R$ there can be at most $k - 1$ points of Y besides the origin in the ball $(2R)S$, while at most l small balls can be accommodated in the ball $(R + r)S$. Also if $y_{\max}$ is the point of Y in W which is farthest from the origin, then $|y_{\max}| > 2R$ and the hyperplane passing through $y_{\max}$ and perpendicular to the line joining the origin to $y_{\max}$ divides the space into two half-spaces such that the ball $(2R)S$ and all the Poisson points of Y in W lie on one half-space H_1 (say) and in the other half-space H_2 (say) there is no point of Y which is at a distance less than $2R$ from $y_{\max}$. This means that there must be a semisphere $T(y_{\max})$ of radius $2R$ centred at $y_{\max}$ and lying in H_2 which contains no points of Y.

To make this formal, we need a conditioning argument. First note that

$$P_\lambda \{\#W = (k, l),\ \mathrm{rad}_Y(W) \geq 2R\}$$

$$\leq \sum_{m=2}^{k+l} P_\lambda (\#W = (k, l),\ mR < \mathrm{rad}_Y(W) \leq (m + 1)R).$$

To estimate the summands, we condition on W_p, the position of the Poisson points of both Y and Z in the ball $(m + 1)RS$. Let $y_{\max}$ be the farthest point of

Y in W_p which is connected to the origin in W_p. From our discussion

$$P_\lambda \{\#W = (k, l), \, mR < \mathrm{rad}_Y(W) \le (m+1)R\}$$

$$\le \sum_{i=0}^{k-1} \sum_{j=0}^{l} E_\lambda(P_\lambda(\{Y(T(y_{\max})\backslash(m+1)RS) = 0\} \cap A_i \cap B_j | W_p)$$

$$= \sum_{i=0}^{k-1} \sum_{j=0}^{l} E_\lambda(P_\lambda(\{Y(T(y_{\max})\backslash(m+1)RS) = 0\} | W_p) 1_{A_i} 1_{B_j})$$

$$\le \exp(-\lambda p \pi_d (R/2)^d) \sum_{i=0}^{k-1} \sum_{j=0}^{l} P_\lambda(A_i) P_\lambda(B_j). \tag{5.37}$$

In the last inequality above we have used the fact that the events A_i and B_j are independent and that the distribution of the Poisson process outside $(m+1)RS$ is independent of W_p.

Now

$$\sum_{i=0}^{k-1} \sum_{j=0}^{l} P_\lambda(A_i) P_\lambda(B_j)$$

$$= \exp(-\lambda \pi_d E(\rho + R)^d)$$

$$\times \sum_{i=0}^{k-1} \sum_{j=0}^{l} \frac{[\lambda p \pi_d (2R)^d]^i}{i!} \frac{[\lambda(1-p)\pi_d(R+r)^d]^j}{j!}. \tag{5.38}$$

Let $c_1 \ge 1$ and $c_2 \ge 1$ be constants such that

$$(2R)^d \le \frac{c_1}{4}\left(\frac{R}{2}\right)^d \quad \text{and} \quad (1-p)(R+r)^d \le \frac{c_2}{4}p\left(\frac{R}{2}\right)^d.$$

With this choice of c_1 and c_2, we see that

$$\sum_{i=0}^{k-1} \sum_{j=0}^{l} \frac{[\lambda p \pi_d (2R)^d]^i}{i!} \frac{[\lambda(1-p)\pi_d(R+r)^d]^j}{j!}$$

$$\le \sum_{i=0}^{k-1} \sum_{j=0}^{l} \frac{[\lambda p \pi_d c_1 \frac{R^d}{2}]^i}{i!} \frac{[\lambda p \pi_d c_2 \frac{R^d}{2}]^j}{j!}$$

$$\le c_1^k c_2^l \sum_{i=0}^{\infty} \sum_{j=0}^{\infty} \frac{[\lambda p \pi_d \frac{R^d}{2}]^i}{i!} \frac{[\lambda p \pi_d \frac{R^d}{2}]^j}{j!}$$

$$= c_1^k c_2^l \exp(\lambda p \pi_d R^d). \tag{5.39}$$

Combining (5.37), (5.38) and (5.39) we obtain (5.35) with appropriate positive constants C_1 and C_2.

The proof of (5.36) is similar and as such we present only an outline of it. Since the origin is the centre of a big ball, if $\#W = (k, l)$ and $\mathrm{rad}_Z(W) > R+r$ there can be at most $l - 1$ small balls centred in $(R + r)S$ and, besides the ball at the origin, at most k big balls centred in $(2R)S$. As in the previous part, we obtain a point $z_{\max}$ and a semisphere $T'(z_{\max})$ of radius $2r$ which is free of Poisson points of Z. Eventually, choosing $c_3 \geq 1$ and $c_4 \geq 1$ to satisfy

$$p(2R)^d \leq \frac{c_3}{4}(1 - p)\left(\frac{r}{2}\right)^d \quad \text{and} \quad (R + r)^d \leq \frac{c_4}{4}\left(\frac{r}{2}\right)^d,$$

we obtain the desired inequality (5.36).   □

Lemma 5.9 *There exist positive constants* C_5, C_6, C_7 *and* C_8 *such that*

$$P_\lambda \{\#W = (k, l), \ R < \mathrm{rad}_Y(W) \leq 2R\}$$
$$\leq \exp(-\lambda \pi_d E(\rho + R)^d - \lambda C_5 \pi_{d-1} R^d + C_6 l) \qquad (5.40)$$

and

$$P_\lambda \{\#W = (k, l), \ R < \mathrm{rad}_Z(W) \leq R + r\}$$
$$\leq \exp(-\lambda \pi_d E(\rho + R)^d - \lambda C_7 \pi_{d-1} r^d + C_8 k). \qquad (5.41)$$

Proof   Throughout this proof, with a slight abuse of notation, we shall write $Y \cap C$ for the set of all points of Y in a region $C \subseteq \mathbb{R}^d$. First observe that if $\mathrm{rad}_Y(W) > R$ then $\mathrm{diam}(Y \cap W) > R$, where diam denotes the diameter of a set as introduced in the previous section. If W_p denotes the position of the Poisson points of both Y and Z in $2RS$, we can find two points y_1 and y_2 in $Y \cap W_p$ which are the farthest apart among all pairs of points in $Y \cap W_p$, and let y_0 be the point in $Y \cap W_p$ which is farthest from the line joining y_1 and y_2. As in the proof of Lemma 5.3, we obtain half-spaces H_1, H_2 and H_0, such that H_1 and H_2 are disjoint and there exist two semispheres of radius $2R$ centred at y_1 and y_2 respectively and also a region in H_0 formed by a semisphere of radius $2R$ and bounded in the slab lying between the half-spaces H_1 and H_2 which are all free of Poisson points of Y. If $d(y_1, y_2) > R$, then (as justified in the proof of Lemma 5.3) the last region described above will have a Lebesgue measure at least $cR\pi_{d-1}(2R)^{d-1}$ for some constant $0 < c < 1$. Moreover, since there is a big ball at the origin, there can be at most l small balls centred in $(R+r)S$.

A conditioning argument as in the previous lemma yields

$$P_\lambda \{ \#W = (k, l), \ R < \mathrm{rad}_Y(W) \leq 2R \}$$

$$\leq \exp(-\lambda \pi_d p(2R)^d) \frac{(\lambda \pi_d p(2R)^d)^k}{k!} \exp(-\lambda c \pi_{d-1} R^d)$$

$$\times \sum_{j=0}^l \exp(-\lambda \pi_d (1-p)(R+r)^d) \frac{[\lambda \pi_d (1-p)(R+r)^d]^j}{j!}.$$

Now choosing positive constants C_5 and C_6 suitably as in the previous lemma, we obtain (5.40).

The proof of (5.41) is similar and we omit it. $\square$

In the next lemma we consider the case when the big balls are centred in the optimal cluster region $(\alpha(k/\lambda))S$ for the big balls where α is as in the proof of Lemma 5.7.

Lemma 5.10   *There is a constant $C_9 > 0$ such that*

$$P_\lambda \left\{ \#W = (k, l), \ \mathrm{rad}_Y(W) \leq \alpha \frac{k}{\lambda}, \ \mathrm{rad}_Z(W) \leq R \right\}$$

$$\leq \exp(-\lambda \pi_d E(\rho + R)^d + (l - (d-1)k) \log \lambda + C_9 k), \quad (5.42)$$

where $\alpha = (p\pi_d(2R)^{d-1})^{-1}$.

Proof

$$P_\lambda \left\{ \#W = (k, l), \ \mathrm{rad}_Y(W) \leq \alpha \frac{k}{\lambda}, \ \mathrm{rad}_Z(W) \leq R \right\}$$

$$\leq P_\lambda \left\{ Y\left(\alpha \frac{k}{\lambda} S\right) = k, \ Y\left((2R)S \backslash \alpha \frac{k}{\lambda} S\right) = 0, \right.$$

$$\left. Z(RS) = l, \ Z((R+r)S \backslash RS) = 0 \right\}$$

$$\leq \exp\left(-\lambda p\pi_d \left(\alpha \frac{k}{\lambda}\right)^d - \lambda(1-p)\pi_d R^d\right) \frac{[\lambda p\pi_d(\alpha \frac{k}{\lambda})^d]^k}{k!} \frac{[\lambda(1-p)\pi_d R^d]^l}{l!}$$

$$\times \exp\left(-\lambda \pi_d \left[p\left((2R)^d - \left(\alpha \frac{k}{\lambda}\right)^d\right)\right.\right.$$

$$\left.\left. + (1-p)((R+r)^d - R^d)\right]\right). \tag{5.43}$$

The lemma follows from (5.43) by an application of Stirling's formula for a suitable C_9. □

To take care of medium-sized clusters, we need a result similar to Lemma 5.3.

Lemma 5.11 *Let* $\mu > 1$ *and define, for* $y > 0$,

$$\psi_\mu(y) := \frac{p\pi_{d-1}(2R)^{d-1} y}{4} - \log(ep\pi_d \mu^d y^d).$$

There exists a constant $C_{10} > 0$ *such that, for* λ *large with* $\mu y(k/\lambda) < 2R$, *we have*

$$P_\lambda \left\{ \#W = (k, l),\ y\frac{k}{\lambda} < \text{rad}_Y(W) \le \mu y\frac{k}{\lambda}\ \text{and}\ \text{rad}_Z(W) \le R \right\}$$

$$\le \exp(-\lambda\pi_d E(\rho + R)^d + (l - (d-1)k)\log\lambda - k\psi_\mu(y) + C_{10}(k)).$$

Proof The proof closely follows the proof of Lemma 5.3. Let

$$\Pi_Y := Y \cap \left(\mu y\frac{k}{\lambda}\right) S, \quad \Pi_Z := Z \cap RS$$

and let

$$A := \left\{ Y \cap \left(\mu\frac{k}{\lambda}\right) S = k \right\}, \quad B := \{Z(RS) = l\}.$$

For $a > 0$ let

$$S_Y(a) := \cup_{y \in (\Pi_Y \setminus \{0\})} S(y, a), \quad S_Z(a) := \cup_{z \in \Pi_Z} S(z, a).$$

If $\mu y(k/\lambda) < 2R$, then all the big balls are in the same component, and so, as in (5.14), we have

$$P_\lambda \left\{ \#W = (k, l),\ y\frac{k}{\lambda} < \text{rad}_Y(W) \le \mu y\frac{k}{\lambda}\ \text{and}\ \text{rad}_Z(W) \le R \right\}$$

$$= E_\lambda \left(P_\lambda \left\{ \#W = (k, l),\ y\frac{k}{\lambda} < \text{rad}_Y(W) \le \mu y\frac{k}{\lambda} \right.\right.$$

$$\text{and}\ \text{rad}_Z(W) \le R | \Pi_Y, \Pi_Z \})$$

$$= E_\lambda \left(1_A 1_B P_\lambda \left\{ Y \left((S_Y(2R) \cup S_Z(R + r)) \setminus \left(\mu y\frac{k}{\lambda}\right) S \right) = 0, \right.\right.$$

$$Z((S_Y(R + r) \cup S_Z(2R)) \setminus RS) = 0 | \Pi_Y, \Pi_Z \right\}). \quad (5.44)$$

Since there is a big ball at the origin, $\ell(S_Y(R+r)\backslash RS) \geq \pi_d((R+r)^d - R^d)$. Also, if $\mathrm{rad}_Y > y(k/\lambda)$, then exactly the same argument as used to justify (5.15) and (5.16) yields

$$\ell\left(S_Y(2R) \backslash \left(\mu y \frac{k}{\lambda}\right) S\right) \geq \pi_d(2R)^d + \pi_{d-1}(2R)^{d-1} y\frac{k}{4\lambda} - \pi_d\left(\mu y \frac{k}{\lambda}\right)^d.$$

Combining this observation with (5.44) and using independence properties we have

$$P_\lambda\left\{\#W = (k, l)\ y\frac{k}{\lambda} < \mathrm{rad}_Y(W) \leq \mu y\frac{k}{\lambda}\ \text{and}\ \mathrm{rad}_Z(W) \leq R\right\} \quad (5.45)$$

$$\leq \exp\left(-\lambda p \pi_d\left(\mu y \frac{k}{\lambda}\right)^d - \lambda(1-p)\pi_d R^d\right)$$

$$\times \frac{[\lambda p \pi_d(\mu y(k/\lambda))^d]^k}{k!} \frac{[\lambda(1-p)\pi_d R^d]^l}{l!}$$

$$\times \exp\left(-\lambda p\left[\pi_d(2R)^d + \pi_{d-1}(2R)^{d-1}y\frac{k}{4\lambda} - \pi_d\left(\mu y \frac{k}{\lambda}\right)^d\right]\right)$$

$$\times \exp(-\lambda\pi_d(1-p)((R+r)^d - R^d))$$

$$\leq \exp(-\lambda\pi_d E(\rho + R)^d + (l - (d-1)k\log\lambda$$

$$- k\psi_\mu(y) + C_{10}k), \quad (5.46)$$

for some positive constant C_{10}, where we have used Stirling's formula in the last inequality. $\qquad\square$

We use the previous lemma to take care of medium-sized components.

Lemma 5.12 *There exists $\beta < \infty$, a positive constant C_{11} and $\lambda_0 < \infty$ such that, for $\lambda \geq \lambda_0$,*

$$P_\lambda\left\{\#W = (k, l),\ \beta\frac{k}{\lambda} < \mathrm{rad}_Y(W) \leq R,\ \mathrm{rad}_Z(W) \leq R\right\}$$

$$\leq \exp(-\lambda\pi_d E(\rho + R)^d + (l - (d-1)k)\log\lambda + C_{11}k).$$

Proof Fix $\mu > 1$ and choose β large such that $\psi_\mu(\mu^j\beta) \geq j$ for every $j \geq 1$. (Note, the function ψ_μ admits such a choice.) Let λ_0 be such that, for all $\lambda \geq \lambda_0$,

$$\beta\mu\frac{k}{\lambda} < r_1 \text{ and, for some } j \geq 1,\ R < \beta\mu^j\frac{k}{\lambda} \leq 2R. \quad (5.47)$$

Let $N := \min\{j : \beta\mu^j k > R\}$. From (5.47), $N \geq 1$. As in Lemma 5.5 we cover the interval $(\beta\mu(k/\lambda), R]$ by intervals of the type $(\beta\mu^{j-1}(k/\lambda), \beta\mu^j(k/\lambda)]$, $j = 0, 1, \ldots, N-1$, and apply Lemma 5.11 to obtain, for $k \geq 1$,

$$P_\lambda \left\{ \#W = (k,l),\ \beta\frac{k}{\lambda} < \mathrm{rad}_Y(W) \leq R,\ \mathrm{rad}_Z \leq R \right\}$$

$$\leq \sum_{j=0}^{N-1} P_\lambda \left\{ \#W = (k,l),\ \beta\mu^j\frac{k}{\lambda} < \mathrm{rad}_Y(W) \leq \beta\mu^{j+1}\frac{k}{\lambda},\ \mathrm{rad}_Z(W) \leq R \right\}$$

$$\leq \sum_{j=0}^{N-1} \exp(-\lambda\pi_d E(\rho + R)^d + (l - (d-1)k)\log\lambda$$

$$\qquad - k\psi_\mu(\beta\mu^j) + C_{10}k)$$

$$\leq \exp(-\lambda\pi_d E(\rho + R)^d + (l - (d-1)k)\log\lambda + C_{10}k)$$

$$\qquad \times \sum_{j=0}^{N-1} \exp(-k\psi_\mu(\beta\mu^j))$$

$$\leq \exp(-\lambda\pi_d E(\rho + R)^d + (l - (d-1)k)\log\lambda + C_{10}k)$$

$$\qquad \times \sum_{j=0}^{\infty} \exp(-kj)$$

$$\leq 2\exp(-\lambda\pi_d E(\rho + R)^d + (l - (d-1)k)\log\lambda + C_{10}k).$$

This proves the lemma. $\qquad\qquad\qquad\qquad\qquad\qquad\qquad\qquad\qquad\qquad\square$

Proof of Theorem 5.4   To begin with, we observe that Lemmas 5.8–5.12 obtain upper bounds for all possible radii of the component W except the case when $\alpha(k/\lambda) < \mathrm{rad}_Y(W) \leq \beta(k/\lambda)$ (if $\alpha < \beta$) and $\mathrm{rad}_Z(W) \leq R$. For this case, we may apply Lemma 5.11 with $\mu = \beta/\alpha > 1$ to obtain

$$P_\lambda \left\{ \#W = (k,l),\ \alpha\frac{k}{\lambda} < \mathrm{rad}_Y(W) \leq \beta\frac{k}{\lambda},\ \mathrm{rad}_Z \leq R \right\}$$

$$\leq \exp(-\lambda\pi_d E(\rho + R)^d + (l - (d-1)k)\log\lambda$$

$$\qquad - k\psi_\mu(\alpha) + C_{12}k), \tag{5.48}$$

where C_{12} is a positive constant depending on μ.

The upper bounds obtained in (5.48), Lemmas 5.8, 5.9, 5.10 and 5.12 together with the lower bound obtained in Lemma 5.7 prove (i) of the theorem.

To prove (ii), we observe that, for $0 < a(\lambda) < R - r$,

$$P_\lambda\{\text{rad}(W) > a(\lambda)|\#W = (k, l)\}$$

$$\leq P_\lambda\{\text{rad}_Z(W) > a(\lambda)|\#W = (k, l)\}. \qquad (5.49)$$

Since a big ball is placed at the origin, if $\text{rad}_Z(W) \leq R$, the small balls may be centred uniformly in $(R - r)S$ without affecting the component. Thus,

$$P_\lambda\{\text{rad}_Z(W) > a(\lambda)|\#W = (k, l), \text{rad}_Z(W) \leq R\} \geq 1 - \left(\frac{a(\lambda)}{R - r}\right)^{dl}. \qquad (5.50)$$

Now choose $a(\lambda)$ such that $a(\lambda) \to 0$ as $\lambda \to \infty$, to obtain, from (5.49), (5.50), (5.36), (5.41) and (i) of the theorem,

$$P_\lambda\{\text{rad}(W) > a(\lambda)|\#W = (k, l)\} \to 1 \text{ as } \lambda \to \infty.$$

Finally,

$$\phi(\lambda) = \frac{k + l + 1}{\lambda \ell(W_H)} \leq \frac{k + l + 1}{\lambda \pi_d (\text{rad}(W))^d}.$$

In (ii) if we choose $a(\lambda)$ such that $a(\lambda) \to 0$ and $\lambda a(\lambda)^d \to \infty$ as $\lambda \to \infty$, then we have, as $\lambda \to \infty$,

$$P_\lambda\left\{\phi(\lambda) \leq \frac{k + l + 1}{\pi_d \lambda a(\lambda)^d}|\#W = (k, l)\right\} \to 1.$$

This proves (iii). □

5.4 Notes

Kertesz and Vicsek (1982) conjectured, based on simulations, that the critical covered volume fraction should be a universal constant for all Poisson Boolean models. In Phani and Dhar (1984) a heuristic argument was given showing that this is not the case, and finally the conjecture was disproved (Theorem 5.1 above) in Meester, Roy and Sarkar (1994). The results in Section 5.2 are due to Alexander (1991), and the results in Section 5.3 are taken from Sarkar (1994).

6

The Poisson random-connection model

A random-connection model (RCM) which is driven by a Poisson process with density λ and connection function g will be denoted by (X, g, λ). In this chapter we will always assume that X has a point at the origin. In Section 1.5, we defined g as a function from $I\!\!R$ into $[0, 1]$, and two points x_1 and x_2 of X are connected to each other with probability $g(|x_1 - x_2|)$. It will be convenient however, to define g as a function from $I\!\!R^d$ into $[0, 1]$ with the following restrictions:

$$g(x) = g(y) \quad \text{whenever } |x| = |y|,$$

and

$$g(x) \le g(y) \quad \text{whenever } |x| \ge |y|.$$

The only reason for this different point of view is that the notation and formulae will be somewhat simpler.

As in Boolean models, some restrictions on g are necessary in order to obtain a non-trivial model, i.e. a model with a non-trivial phase transition. The first section of this chapter is devoted to that problem. In the second section, we shall derive some useful but technical results concerning the connection function g. In the third section, we shall demonstrate that unlike the Boolean model, equality of the two most important critical densities is always true here. Further topics in this chapter are uniqueness and high-density processes.

Unfortunately, the proofs in the RCM tend to be quite technical. To some extent, this is the price we have to pay for allowing a large class of connection functions. The ideas behind the various proofs, however, are very often not so hard to grasp, and we shall always try to give the reader an idea of what is going on. Some words about the notation: we denote by $W(x)$ the component containing the point x of X, and W denotes the component containing the origin. The cardinality of a component W is denoted by $|W|$. The origin itself is as earlier denoted by 0. The probability measure in the model will be denoted by $P_{(\lambda,g)} = P_\lambda = P$, when no confusion is possible and the corresponding expectation operator will be denoted by $E_{(\lambda,g)} = E_\lambda = E$.

6.1 Non-triviality of the model

It is not hard to see that for certain functions g, no finite components exist a.s. For this, we just take a look at Proposition 1.3. If Y denotes the (random) number of points of X which are connected directly to the origin, then we see that

$$P(Y = k) = e^{-\lambda \int_{\mathbb{R}^d} g(x)\, dx} \frac{(\lambda \int_{\mathbb{R}^d} g(x)\, dx)^k}{k!}. \tag{6.1}$$

Hence, if $\int_{\mathbb{R}^d} g(x)\, dx = \infty$, then $P(Y = k) = 0$ for all k and the conclusion is that $Y = \infty$ a.s. This is true for all $\lambda > 0$, and it implies that percolation occurs for all positive values of λ. So in order for the model to become more interesting, we need the condition

$$0 < \int_{\mathbb{R}^d} g(x)\, dx < \infty. \tag{6.2}$$

Our first result shows that this necessary condition is also sufficient for the occurrence of a non-trivial phase transition. We write $\theta_g(\lambda) = \theta(\lambda) = P_{(\lambda,g)}(|W| = \infty)$ and $\chi(\lambda) = \chi_g(\lambda) = E_{(\lambda,g)}(|W|)$.

Theorem 6.1 *Consider a Poisson RCM (X, g, λ) in $\mathbb{R}^d$, for $d \geq 2$. If g satisfies (6.2), then there exist two densities $0 < \lambda_T(g) \leq \lambda_H(g) < \infty$ such that*

(1) $\chi(\lambda) < \infty$ *for* $\lambda < \lambda_T(g)$, *and* $\chi(\lambda) = \infty$ *for* $\lambda > \lambda_T(g)$.
(2) $\theta(\lambda) = 0$ *for* $\lambda < \lambda_H(g)$ *and* $\theta(\lambda) > 0$ *for* $\lambda > \lambda_H(g)$.

Proof Using coupling as before, we see that both χ and θ are non-decreasing in λ, whence it is clear that $\lambda_T(g)$ and $\lambda_H(g)$ with the properties in (1) and (2) exist and that $\lambda_T(g) \leq \lambda_H(g)$. It suffices therefore to show that $\lambda_T(g) > 0$ and that $\lambda_H(g) < \infty$.

We need to show then, that for λ sufficiently small (but positive) the expected size of W is finite. We tackle this problem with a branching process argument together with coupling. The branching process argument is quite similar to that in Chapter 3. As we saw in (6.1), the expected number of points connected directly to the origin is equal to $\lambda \int_{\mathbb{R}^d} g(x)\, dx$. More precisely, the points connected directly to the origin form a non-homogeneous Poisson process with intensity function $\lambda g(x)$. We denote this point process by X_0. We are going to 'build' a

Poisson process as the superposition of many non-homogeneous Poisson processes as follows. Suppose that the points of X_0 are given by $x_1, x_2, \ldots, x_n$. We call this the points of the first generation. To construct the second generation we proceed as follows. Take the first point, x_1, of the first generation, and superpose it with a non-homogeneous Poisson process X_1^1 with intensity function $\lambda(1 - g(x))g(x - x_1)$ and which is independent of X_0. The occurrences of X_1^1 are the second generation points coming from x_1 and they represent all points which are connected to x_1 (and possibly to $x_2, \ldots, x_n$) but not to the origin. For x_2, we take another non-homogeneous Poisson process X_1^2 with intensity function $\lambda(1-g(x))(1-g(x-x_1))g(x-x_2)$, independent of X_0 and X_1^1. These are the points of the second generation coming from x_2 and they represent all points which are connected to x_2 but not to the origin and x_1. We continue this procedure in the obvious way, obtaining non-homogeneous Poisson processes $X_1^1, \ldots, X_1^n$ which are all independent of each other. For each i, the intensity function of X_1^i contains the factor $g(x - x_i)$ and therefore X_1^i can be coupled to independent non-homogeneous Poisson processes $\tilde{X}_1^i$ with intensity functions $\lambda g(x - x_i)$ such that the occurrences of X_1^i are a subset of the occurrences of $\tilde{X}_1^i$. The total number of points of $\tilde{X}_1^i$ is a random variable with a Poisson distribution with parameter $\lambda \int_{\mathbb{R}^d} g(x - x_i)\, dx = \lambda \int_{\mathbb{R}^d} g(x)\, dx$. Hence, in the coupling just described, the total number of points in the second generation is bounded from above by the total number of points in the second generation of an ordinary Galton–Watson branching process with expected offspring equal to $\lambda \int_{\mathbb{R}^d} g(x)\, dx$. In general, the number of points in the n-th generation is bounded from above by the number of points in the n-th generation of such a branching process. It is well known (see e.g. Grimmett and Stirzaker 1992, Lemma 5.4.2) that the expected number of points in the n-th generation of an ordinary Galton–Watson branching process with expected offspring μ is equal to μ^n. Hence, the expected number of points in W satisfies

$$E_\lambda(|W|) \le \sum_{n=1}^{\infty} \left(\lambda \int_{\mathbb{R}^d} g(x)\, dx \right)^n. \qquad (6.3)$$

Thus, if g satisfies (6.2) we can choose $\lambda < \left(\int_{\mathbb{R}^d} g(x)\, dx \right)^{-1}$ to make the sum in (6.3) finite. This shows that $\lambda_T(g) \ge \left(\int_{\mathbb{R}^d} g(x)\, dx \right)^{-1} > 0$.

For the second part of the theorem we need to show that for λ sufficiently large, we have $|W| = \infty$ with positive probability. For this, we use the notion of the *Lebesgue set* of a function g. This is defined as the set of points $y \in \mathbb{R}^d$ such that

$$\lim_{\epsilon \downarrow 0} (2\epsilon)^{-d} \int_{y + B_\epsilon} |g(x) - g(y_0)|\, dx = 0, \qquad (6.4)$$

where $B_\epsilon = [-\epsilon, \epsilon]^d$ and $y + B_\epsilon$ denotes the set $\{y + x : x \in B_\epsilon\}$. It is well known (see Rudin 1970, Theorem 8.8) that the Lebesgue set of g has full Lebesgue measure. Hence we can select d linearly independent points $y_1, \ldots, y_d$ which are all in the Lebesgue set of g and for which $g(y_i) > 0, i = 1, \ldots, d$. From (6.4) it follows that we can find $\delta > 0$ such that for all i and for all boxes B with side length at most δ containing some y_i,

$$\int_B g(x)\, dx \geq \tfrac{1}{2}\ell(B)g(y_i), \tag{6.5}$$

for all i. At the same time, δ can be taken so small that all sets of the form $n_1 y_1 + \cdots + n_d y_d + B_\delta =: B_\delta(n_1, \ldots, n_d), n_1, \ldots, n_d \in \mathbf{Z}$, are mutually disjoint.

For any edge e in $\mathbf{Z}^d$ between vertices $(n_1, \ldots, n_d)$ and $(m_1, \ldots, m_d)$ with $\sum_{i=1}^d |n_i - m_i| = 1$, we place an independent Poisson process X_e with density $(2d)^{-1}\lambda$ on the boxes $B_\delta(n_1, \ldots, n_d)$ and $B_\delta(m_1, \ldots, m_d)$. On the complement of the union of all these sets, we place a Poisson process X^* with density λ, independent of all other processes. The superposition of X^* with X_e for all edges e yields a homogeneous Poisson process X with density λ on $\mathbb{R}^d$.

Let $n = (n_1, \ldots, n_d)$ and $m = (m_1, \ldots, m_d)$ be such that $\sum_{i=1}^d |n_i - m_i| = 1$ and let e be the edge between n and m. Then $B_\delta(m)$ can be written as $B_\delta(n) + y_i$ for suitable i. Given a point x of X in the box $B_\delta(n)$, the probability that x is not connected to any point of X_e in $B_\delta(m)$ is then equal to

$$\exp\left(-\lambda(2d)^{-1} \int_{B_\delta} g(y - x)\, dy\right)$$
$$\leq \exp\left(-\lambda(2d)^{-1}\tfrac{1}{2}(2\epsilon)^d g(y_i)\right)$$
$$= \exp\left(-\lambda(4d)^{-1}(2\epsilon)^d g(y_i)\right), \tag{6.6}$$

where $B_\delta = B_\delta(m_1, \ldots, m_d)$ and where we have used (6.5). Now we can perform independent bond percolation on $\mathbf{Z}^d$ as follows. The cluster C of the origin in this discrete percolation model is built in steps. We start with C consisting of the origin only, and define x_0 to be the origin. In an inductive fashion, assume that C consists of a finite number of vertices such that for each $n \in C$, we have chosen a point x_n of X inside the box $B_\delta(n)$. Next we consider an edge e between n and m which has not been considered before, such that n belongs to C, but m does not. If no such edge exists we stop, but if it exists we check whether or not there is a point of X_e in $B_\delta(m)$ which is connected to x_n. If such a point exists we choose one, denote it by x_m and add m to C.

It follows from the construction and (6.6) that the cluster C obtained by this procedure can be seen as the cluster of the origin in discrete bond percolation with parameter at least $\min_{1 \le i \le d}\{1 - e^{-\lambda(4d)^{-1}(2\epsilon)^d g(y_i)}\}$. Hence for λ sufficiently large, the probability that the cluster of the origin is infinite (which means that the inductive procedure above does not stop) is positive. But if C is infinite, then certainly the component W in the underlying RCM is unbounded. This observation completes the proof of the theorem. □

6.2 Properties of the connection function

In this section we collect some technical results which will be used later. At this stage, the idea behind the various definitions might not be so clear, but the readability of the subsequent section certainly increases if we isolate all technical lemmas. Readers not interested in technical details may just note the statements of the results and move on to the next section.

First, we define three functions based on the connection function g.

Definition 6.1 *For $L > 0$, the function $g_L : \mathbb{R}^d \to [0, 1]$ is defined as*

$$g_L(x) = 1 - \prod_{z \in \mathbb{Z}^d} (1 - g(x + 2Lz)). \tag{6.7}$$

Definition 6.2 *For $y, x_1, \ldots, x_k \in \mathbb{R}^d$, let $g_1(y; x_1, \ldots, x_k)$ be the probability that in the random graph with (non-random) vertices $y, x_1, \ldots, x_k$ and connection function g, the point y is not isolated. Furthermore, $g_2(x_1, \ldots, x_k)$ is defined to be the probability that the graph with vertices $x_1, \ldots, x_k$ and connection function g is connected.*

It is clear from the definition that $g_L(x) \ge g(x)$ for all L and x. Here are some further properties of g and g_L:

Proposition 6.1

   (i) *For every L, the function g_L is continuous almost everywhere (with respect to Lebesgue measure).*

   (ii) *For all $\epsilon > 0$, we have, for all L large enough, $|g_L(x) - g(x)| < \epsilon$, for all $x \in B_L$.*

   (iii) $\lim_{L \to \infty} \int_{B_L} g_L(x)\, dx = \int_{\mathbb{R}^d} g(x)\, dx$.

Proof   (i) As g is a non-increasing function of the absolute value, its set of discontinuities has Lebesgue measure zero. Hence, we can restrict our attention to those x for which g is continuous at all points of the form $x + 2Lz, z \in \mathbf{Z}^d$.

First suppose x is such that $g(x + 2Lz) = 1$ for some $z \in \mathbf{Z}^d$. Then of course $g_L(x) = 1$. For any sequence $\{x_n\}$ converging to x we have that $g(x_n + 2Lz) \to 1$ and it follows that also $g_L(x_n) \to 1$.

Now suppose $g(x + 2Lz) < 1$ for all z. This implies that $g(x + 2Lz)$ is bounded away from 1. Furthermore, if $\{x_n\}$ converges to x, it follows from the continuity of g at all points $x + 2Lz$ and the fact that $g(z)$ tends to zero whenever $|z|$ tends to infinity, that $g(x_n + 2Lz)$ is uniformly bounded away from 1, for all z and for all n large enough. We now have $-\log(1 - g_L(x)) = \sum_{z \in \mathbf{Z}^d} -\log(1 - g(x + 2Lz))$. It is enough to show that we can interchange limit and sum in the expression $\lim_{n \to \infty} \sum_{z \in \mathbf{Z}^d} -\log(1 - g(x_n + 2Lz))$. This is not hard, we write, for large K:

$$\sum_{|z|>K} -\log(1 - g(x_n + 2Lz))$$

$$= \sum_{|z|>K} \log((1 - g(x_n + 2Lz))^{-1})$$

$$= \sum_{|z|>K} \log\left(1 + \sum_{k=1}^{\infty} (g(x_n + 2Lz))^k\right).$$

This is bounded from above by

$$\sum_{|z|>K} \frac{g(x_n + 2Lz)}{1 - g(x_n + 2Lz)} \leq C \sum_{|z|>K} g(x_n + 2Lz),$$

for some constant C. This tends to zero uniformly in n when $K \to \infty$, using condition (6.2).

For (ii), let $x \in B_L$ be arbitrary. For any $\epsilon > 0$, we can take L large enough so that

$$\prod_{z \in \mathbf{Z}^d \setminus 0} (1 - g(x + 2Lz))$$

$$\geq 1 - \sum_{z \in \mathbf{Z}^d \setminus 0} g(x + 2Lz)$$

$$\geq 1 - \int_{\mathbb{R}^d \setminus B_{L-1}} g(y)\, dy \geq 1 - \epsilon.$$

(Here we use the integrability condition (6.2).) For such L, we thus find that $1 - g_L(x) \geq (1 - \epsilon)(1 - g(x))$, which implies that $g_L(x) - g(x) \leq \epsilon$. As $g_L(x) \geq g(x)$ for all x, the result follows.

For (iii) we write

$$g_L(x) = 1 - \prod_{z \in \mathbb{Z}^d} (1 - g(x + 2Lz)) \le \sum_{z \in \mathbb{Z}^d} g(x + 2Lz).$$

Thus, on the one hand,

$$\int_{B_L} g_L(x)\, dx \le \int_{B_L} \sum_{z \in \mathbb{Z}^d} g(x + 2Lz)\, dx$$

$$= \sum_{z \in \mathbb{Z}^d} \int_{B_L} g(x + 2Lz)\, dx$$

$$= \int_{\mathbb{R}^d} g(x)\, dx.$$

On the other hand, $\int_{B_L} g_L(x) dx \ge \int_{B_L} g(x) dx$, which converges to $\int_{\mathbb{R}^d} g(x) dx$ when $L \to \infty$. $\qquad\square$

The next result explains why g_1 and g_2 are defined the way they are:

Proposition 6.2 *Consider a Poisson RCM (X, g, λ) in $\mathbb{R}^d$. It is the case that*

$$P(|W| = k, W \subset B_n) = \frac{\lambda^{k-1}}{(k-1)!} \int_{B_n \times \cdots \times B_n} g_2(0, x_1, \ldots, x_{k-1})$$

$$\times \exp\left(-\lambda \int_{\mathbb{R}^d} g_1(y; 0, x_1, \ldots, x_{k-1})\, dy\right) d(x_1, \ldots, x_{k-1}). \quad (6.8)$$

Proof We denote by $E(k, n)$ the event that $|W| = k$ and $W \subset B_n$. Conditioned on the event that $X(B_n) = m$, for $m \ge k - 1$, we know from Proposition 1.2 that the m points of X in B_n are uniformly distributed on B_n. This implies that (remember that the origin always belongs to W)

$$P(E(k, n) \mid X(B_n) = m) = \binom{m}{k-1} (2n)^{-dm}$$

$$\times \int_{B_n} \cdots \int_{B_n} P'(W = \{0, x_1, \ldots, x_{k-1}\})\, dx_1 \cdots dx_m, \quad (6.9)$$

where P' denotes the probability measure of a RCM where we superpose the origin 0 and the points $x_1, \ldots, x_{m-1}$ with a Poisson process with density λ on the complement of B_n, and connect any two points according to the connection function g as usual.

The probability that a point y is connected to at least one of the points $\{0, x_1, \ldots, x_{k-1}\}$ is $g_1(y; 0, x_1, \ldots x_{k-1})$ by definition. Hence the probability that no point of the Poisson process with density λ on the complement of B_n is connected to any of the points $0, x_1, \ldots, x_k$ is equal to

$$\exp\left(-\lambda \int_{\mathbb{R}^d \setminus B_n} g_1(y; 0, x_1, \ldots, x_{k-1}) \, dy\right).$$

From the definition of g_2 we then obtain

$$P'(W = \{0, x_1, \ldots, x_{k-1}\}) = g_2(0, x_1, \ldots, x_{k-1})$$
$$\times \exp\left(-\lambda \int_{\mathbb{R}^d \setminus B_n} g_1(y; 0, x_1, \ldots, x_{k-1}) \, dy\right)$$
$$\times \prod_{i=k}^{m}(1 - g_1(x_i; 0, x_1, \ldots, x_{k-1})).$$

This we substitute into (6.9) to obtain

$$P(E(k, n) \cap \{X(B_n) = m\}) = e^{-\lambda(2n)^d} \lambda^m ((k-1)!(m-(k-1))!)^{-1}$$
$$\times \int_{B_n} \cdots \int_{B_n} g_2(0, x_1, \ldots, x_{k-1})$$
$$\times \exp\left(-\lambda \int_{\mathbb{R}^d \setminus B_n} g_1(y; 0, x_1, \ldots, x_{k-1}) \, dy\right)$$
$$\times \left(\int_{B_n}(1 - g_1(z; 0, x_1, \ldots, x_{k-1})) \, dz\right)^{m-(k-1)} dx_1 \cdots dx_{k-1}.$$

Summing this last expression over $m \geq k - 1$ yields

$$P(E(k, n)) = \frac{e^{-\lambda(2n)^d} \lambda^{k-1}}{(k-1)!} \int_{B_n} \cdots \int_{B_n} g_2(0, x_1, \ldots, x_{k-1})$$
$$\times \exp\left\{\lambda \int_{B_n}(1 - g_1(z; 0, x_1, \ldots, x_{k-1})) \, dz\right.$$
$$\left. -\lambda \int_{\mathbb{R}^d \setminus B_n} g_1(y; 0, x_1, \ldots, x_{k-1}) \, dy\right\} dx_1 \cdots dx_{k-1}$$
$$= \frac{e^{-\lambda(2n)^d} \lambda^{k-1}}{(k-1)!} \int_{B_n} \cdots \int_{B_n} g_2(0, x_1, \ldots, x_{k-1})$$
$$\times \exp\left\{-\lambda \int_{\mathbb{R}^d} g_1(y; 0, x_1, \ldots, x_{k-1}) \, dy + \lambda \int_{B_n} dz\right\}$$
$$dx_1 \cdots dx_{k-1}$$

$$= \frac{\lambda^{k-1}}{(k-1)!} \int_{B_n} \cdots \int_{B_n} g_2(0, x_1, \ldots, x_{k-1})$$

$$\times \exp\left\{-\lambda \int_{\mathbb{R}^d} g_1(y; 0, x_1, \ldots, x_{k-1}) \, dy\right\} dx_1 \cdots dx_{k-1}.$$

$\square$

Finally, we shall need the following result:

Proposition 6.3 *Suppose that g has bounded support. Then we have*

$$\liminf_{|h| \to 0} |h|^{-1} \int_{\mathbb{R}^d} |g(x + h) - g(x)| \, dx > 0. \tag{6.10}$$

Proof It is easy to see that

$$\int_{\mathbb{R}^d} (g(x - h) - g(x)) x \, dx = h \int_{\mathbb{R}^d} g(x) \, dx.$$

Hence it follows that

$$\int_{\mathbb{R}^d} |x| \cdot |g(x - h) - g(x)| \, dx \geq |h| \int_{\mathbb{R}^d} g(x) \, dx.$$

We can find a number $r > 0$ such that $g(x) = 0$, whenever $|x| > r$. If $|h| \leq 1$, then both $g(x - h)$ and $g(x)$ are zero for x satisfying $|x| > r + 1$. It follows that for h with $|h| \leq 1$ we have

$$|h| \int_{\mathbb{R}^d} g(x) \, dx \leq (r + 1) \int_{\mathbb{R}^d} |g(x - h) - g(x)| \, dx,$$

which implies the desired result because $\int_{\mathbb{R}^d} g(x) \, dx > 0$. $\square$

6.3 Equality of the critical densities

The critical densities $\lambda_H(g)$ and $\lambda_T(g)$ defined in Section 6.1 satisfy the obvious inequality $\lambda_T(g) \leq \lambda_H(g)$. It is a very natural question as to whether or not these densities are actually the same. In Chapter 3 we proved that this need not be the case in a Boolean model. The reason for this latter fact is that one ball can give rise to a very large volume. This phenomenon does not occur in random connection models and we can prove the following result:

Theorem 6.2 *For every connection function g we have*

$$\lambda_H(g) = \lambda_T(g).$$

As far as condition (6.2) is concerned, note that if $\int_{\mathbb{R}^d} g(x)\,dx = \infty$, then $\lambda_H(g) = \lambda_T(g) = 0$ and in case $\int_{\mathbb{R}^d} g(x)\,dx = 0$, then $\lambda_H(g) = \lambda_T(g) = \infty$. Note that Theorem 6.2 implies the corresponding statement in Boolean models with fixed-size balls, i.e. in the case where $g(x) = I_{\{|x| \le r\}}$ for some positive r. As in the proof for the Boolean model, the proof which we shall give here uses discrete approximation of the model. It is easy to get completely lost in the proof of Theorem 6.2, so it pays to take a moment to explain the strategy of the proof.

The first step of the proof is to introduce an extra parameter $0 < \gamma < 1$ in the model, in such a way that the original model can in some sense be viewed as the limit for $\gamma \to 0$. In the second step, the new model (with the extra parameter γ) is then approximated by a discrete percolation model in a finite box. Here we encounter a difficulty. It is necessary to have a notion of stationarity also in the model in a finite box. This is not automatically the case, because different points have different positions with respect to the boundary of the box. Hence we adapt our model to this end. It is here where we use the map g_L defined previously. In the third step, we derive two fundamental differential inequalities governing the behaviour of the important quantities in the discrete model in the finite box. The next two steps consist of limit procedures: one to go from the discrete model at finite volume to the continuum model at finite volume, and the other from the continuum model at finite volume to the continuum model on $\mathbb{R}^d$. Finally, we are left with two differential inequalities in the model with the extra parameter γ and then it is not hard to prove that these inequalities imply the desired result.

STEP 1:   Let us start then with the introduction of the extra parameter γ. Consider a realisation of the point process X. We label each point of X with probability γ, where γ is assumed to be strictly between 0 and 1. The (random) set of labelled points is denoted by G. The idea behind the labelled points is to see them as surrogates for 'the point at infinity'. If we denote by $\theta(\lambda, \gamma)$ the probability that the component of the origin W contains a labelled point, then by taking γ smaller and smaller, it is likely that W should be larger and larger in order to contain a labelled point. In a similar fashion, $\chi(\lambda, \gamma)$ denotes the expectation of $|W|$ on the event that W does not contain a labelled point: $\chi(\lambda, \gamma) = E(|W| \cdot 1_{\{W \cap G = \emptyset\}})$. It is in the following sense that the original model is retrieved by taking the limit for $\gamma \to 0$.

Lemma 6.1

    (i) $\lim_{\gamma \to 0} \theta(\lambda, \gamma) = \theta(\lambda)$,

    (ii) $\lim_{\gamma \to 0} \chi(\lambda, \gamma) = E(|W| \cdot 1_{\{|W| < \infty\}})$.

Proof If $|W| = \infty$, then with probability one W contains at least one labelled point. The labelling is independent of the percolation structure and hence we can write

$$\theta(\lambda, \gamma) = 1 - \sum_{n=1}^{\infty} P(W \cap G = \emptyset \mid |W| = n) P(|W| = n)$$

$$= 1 - \sum_{n=1}^{\infty} (1 - \gamma)^n P(|W| = n).$$

This is a power series in $(1 - \gamma)$ with radius of convergence at least 1, and we can take the limit for $\gamma \to 0$ to obtain (i). For the second result, we write

$$\chi(\lambda, \gamma) = \sum_{n=1}^{\infty} n P(W \cap G = \emptyset \mid |W| = n) P(|W| = n)$$

$$= \sum_{n=1}^{\infty} n(1 - \gamma)^n P(|W| = n).$$

Taking the limit for $\gamma \to 0$ gives (ii). $\square$

It is therefore natural to define $\theta(\lambda, 0) = \theta(\lambda)$ and $\chi(\lambda, 0) = \chi^f(\lambda)$, where $\chi^f(\lambda)$ is the expected size of the component of the origin on the event that it is finite.

STEP 2: We continue with the second part of our programme, the approximation of the model by a discrete percolation model in a finite box. One important feature of this discrete model is that both vertices and edges are randomly chosen to be either open or closed. In order to define the approximating models, we choose two parameters L and n, both integers. Let B_L be the box $[-L, L]^d$ and divide this box into little boxes of side length 2^{-n}. Put a vertex in the middle of each of these boxes. A vertex v is said to be open if the Poisson process has at least one point in the small box containing v; otherwise v is said to be closed. Note that the state of a vertex is independent of the states of all other vertices. Next we consider connections between these vertices. We want to obtain a notion of stationarity at finite volume. In order to achieve this, we use the connection function g_L rather than g. Recall that g_L is defined as follows:

$$g_L(x) = 1 - \prod_{z \in \mathbb{Z}^d} (1 - g(x + 2Lz)). \tag{6.11}$$

Note that g_L is translation invariant in the box B_L, see Figure 6.1. We connect any two vertices v and v' in B_L with probability $g_L(v' - v)$, independently of anything else. When v and v' are connected, we say that the edge between them

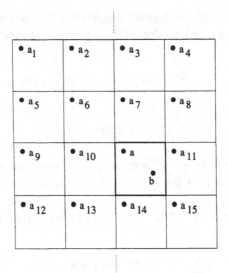

Figure 6.1. The bold line square is B_L. The probability that a and b are connected with the connection function g_L is the probability that b is connected to any of the points $a_1, a_2, \ldots$ when the connection function is g.

is open; otherwise it is closed. Note that an open edge can have zero, one or two closed end points. Also note that the probability for a vertex to be open depends on the density of the Poisson process, but the probability for an edge to be open is independent of the Poisson process (and hence also independent of λ).

It will be convenient to define $C_L^n(v)$, the component of a vertex v, as the set of all vertices v' in B_L for which there exists an alternating sequence ($v_0 = v, e_1, v_1, e_2, v_2, \ldots, e_n, v_n = v'$) of vertices and edges such that e_n connects v_{n-1} and v_n and which are all open, *except* possibly v itself. According to this definition, $v \in C_L^n(v)$ whether v is open or not. In this discrete model, we again label each vertex with probability γ, independently of all other vertices. Note that closed vertices may also be labelled. The component of the origin is denoted by C_L^n, the set of labelled sites by G_L^n, and the relevant functions are

$$\theta_L^n(\lambda, \gamma) = P(C_L^n \cap G_L^n \neq \emptyset), \tag{6.12}$$

$$\chi_L^n(\lambda, \gamma) = E(|C_L^n| \cdot 1_{\{C_L^n \cap G_L^n = \emptyset\}}). \tag{6.13}$$

STEP 3: The third step in the proof is to derive the two differential inequalities.

Lemma 6.2 *Let M_L^n denote the expected number of open edges with one end point at the origin. For $\lambda > 0$ and $0 < \gamma < 1$ it is the case that*

(i) $\dfrac{\partial \theta_L^n}{\partial \lambda} \leq 2^{-nd} \theta_L^n \cdot \chi_L^n \cdot M_L^n,$

(ii) $\theta_L^n \leq \gamma \dfrac{\partial \theta_L^n}{\partial \gamma} + (\theta_L^n)^2 + 2^{nd} (e^{\lambda 2^{-nd}} - 1) \theta_L^n \dfrac{\partial \theta_L^n}{\partial \lambda}.$

Proof For (i), first note that the labelling procedure is independent of every-thing else and hence, given a set of vertices Γ in B_L,

$$P(C_L^n \cap \Gamma \neq \emptyset \mid G_L^n = \Gamma) = P(C_L^n \cap \Gamma \neq \emptyset),$$

whence

$$\theta_L^n(\lambda, \gamma) = \sum_\Gamma P(G_L^n = \Gamma) P(C_L^n \cap \Gamma \neq \emptyset),$$

where the sum is over all possible subsets Γ of vertices in B_L. But $P(G_L^n = \Gamma)$ does not depend on λ and hence, writing $A_L^n(\Gamma)$ for $\{C_L^n \cap \Gamma \neq \emptyset\}$ we obtain

$$\frac{\partial \theta_L^n}{\partial \lambda} = \sum_\Gamma P_\lambda(G_L^n = \Gamma) \frac{d}{d\lambda} P_\lambda(A_L^n(\Gamma)).$$

Next we use Russo's formula Theorem 1.8. The probability for a vertex to be open is equal to $1 - \exp(-\lambda 2^{-nd})$, while the probability for an edge to be open is independent of λ. Hence we obtain

$$\frac{\partial \theta_L^n}{\partial \lambda} = 2^{-nd} \sum_\Gamma P(G_L^n = \Gamma) \sum_v P(v \text{ is pivotal for } A_L^n(\Gamma) \text{ and closed}).$$

$$(6.14)$$

Let the *closure* $cl(C_L^n)$ of C_L^n be the set of vertices and edges consisting of

(i) all vertices in C_L^n,

(ii) all edges (open and closed) with at least one end point in C_L^n,

(iii) all closed neighbours of C_L^n, where two vertices are said to be neighbours if the edge between them is open.

Note that conditioned on $\{C_L^n = \Sigma, cl(C_L^n) = \Sigma^*\}$, the configuration of open and closed vertices and edges outside $cl(C_L^n)$ is still unconditioned and chosen according to the appropriate product measure. Now let $\{C_L^n = \Sigma, cl(C_L^n) = \Sigma^*\} =: E(\Sigma, \Sigma^*)$, where Σ and Σ^* are such that $0 \in \Sigma$, $v \notin \Sigma$ but $v \in \Sigma^*$. Then

$$P(v \text{ is pivotal for } A_L^n(\Gamma) \text{ and closed})$$

$$= \sum_{(\Sigma, \Sigma^*)} P(C_L^n \cap \Gamma = \emptyset, C_{L/\Sigma^*}^n(v) \cap \Gamma \neq \emptyset, E(\Sigma, \Sigma^*)),$$

where $C^n_{L/\Sigma^*}(v)$ is the cluster of v if we restrict ourselves to the graph where we first delete all vertices (other than v) and edges in Σ^* and the sum is over all $\Sigma \subseteq \Sigma^*$ such that $0 \in \Sigma$, $v \notin \Sigma$, $v \in \Sigma^*$. Hence, $\partial \theta^n_L / \partial \lambda$ can be written as

$$2^{-nd} \sum_v \sum_{(\Sigma, \Sigma^*)} \sum_\Gamma P(C^n_L \cap \Gamma = \emptyset, C^n_{L/\Sigma^*}(v) \cap \Gamma \neq \emptyset,$$

$$E(\Sigma, \Sigma^*)) P(G^n_L = \Gamma)$$

$$= 2^{-nd} \sum_v \sum_{(\Sigma, \Sigma^*)} \sum_\Gamma P(C^n_L \cap \Gamma = \emptyset, C^n_{L/\Sigma^*}(v) \cap \Gamma \neq \emptyset,$$

$$E(\Sigma, \Sigma^*), G^n_L = \Gamma)$$

$$= 2^{-nd} \sum_v \sum_{(\Sigma, \Sigma^*)} P(C^n_L \cap G^n_L = \emptyset, C^n_{L/\Sigma^*}(v) \cap G^n_L \neq \emptyset,$$

$$E(\Sigma, \Sigma^*))$$

$$= 2^{-nd} \sum_v \sum_{(\Sigma, \Sigma^*)} P(C^n_L \cap G^n_L = \emptyset, C^n_{L/\Sigma^*}(v) \cap G^n_L \neq \emptyset \mid$$

$$E(\Sigma, \Sigma^*)) P(E(\Sigma, \Sigma^*)).$$

Given the event $E(\Sigma, \Sigma^*)$, the event $\{C^n_L \cap G^n_L = \emptyset\}$ is independent of the event $\{C^n_{L/\Sigma^*}(v) \cap G^n_L \neq \emptyset\}$. Hence we obtain

$$\frac{\partial \theta^n_L}{\partial \lambda} = 2^{-nd} \sum_v \sum_{(\Sigma, \Sigma^*)} P(C^n_L \cap G^n_L = \emptyset \mid E(\Sigma, \Sigma^*)) P(E(\Sigma, \Sigma^*))$$

$$\times P(C^n_{L/\Sigma^*}(v) \cap G^n_L \neq \emptyset \mid E(\Sigma, \Sigma^*))$$

$$\leq 2^{-nd} \sum_v \theta^n_L P(C^n_L \cap G^n_L = \emptyset, v \text{ is a closed neighbour of } C^n_L)$$

$$\leq 2^{-nd} \theta^n_L \sum_v \sum_{v' \neq v} P(C^n_L \cap G^n_L = \emptyset, v' \in C^n_L,$$

$$v \text{ is a closed neighbour of } v')$$

$$= 2^{-nd} \theta^n_L \sum_v \sum_{v' \neq v} P(C^n_L \cap G^n_L = \emptyset, v' \in C^n_L, v \text{ is closed})$$

$$\times P(v \text{ is a neighbour of } v')$$

$$\leq 2^{-nd} \theta^n_L \sum_v \sum_{v' \neq v} P(C^n_L \cap G^n_L = \emptyset, v' \in C^n_L)$$

$$\times P(v \text{ is a neighbour of } v')$$

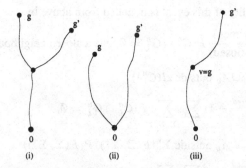

Figure 6.2. Some situations concerning double connectedness. The vertices g and g' are in $C_L^n \cap G_L^n$. In (i), v is doubly connected to G_L^n, in (ii), 0 is doubly connected and in (iii), v is doubly connected in such a way that one of the paths consists of v only.

$$= 2^{-nd} \theta_L^n \sum_{v'} P(C_L^n \cap G_L^n = \emptyset, v' \in C_L^n)$$

$$\times \sum_{v \neq v'} E\left(1_{\{v \text{ is a neighbour of } v'\}}\right)$$

$$= 2^{-nd} \theta_L^n \cdot \chi_L^n \cdot M_L^n.$$

The proof of (ii) is based on the BK-inequality. We write θ_L^n as follows:

$$\theta_L^n = P(|C_L^n \cap G_L^n| = 1) + P(|C_L^n \cap G_L^n| \geq 2). \tag{6.15}$$

The first term in (6.15) is easily computed:

$$P(|C_L^n \cap G_L^n| = 1) = \sum_{k=1}^{\infty} k\gamma(1-\gamma)^{k-1} P(|C_L^n| = k) = \frac{\gamma}{1-\gamma} \chi_L^n = \gamma \frac{\partial \theta_L^n}{\partial \gamma},$$

where the last equality is an easy consequence of the definitions. If $|C_L^n \cap G_L^n| \geq 2$, then it is not hard to see (but quite hard to prove!) that

(i) There exist two edge/site disjoint paths (apart from the origin) connecting the origin to two vertices in G_L^n. (We say that the origin is *doubly connected* to G_L^n.)

(ii) There exists an open vertex v such that if we close v, $C_L^n \cap G_L^n$ becomes empty and v is doubly connected to G_L^n using no vertices in C_L^n.

We refrain from proving this assertion and refer to Figure 6.2 instead. Hence the second term in (6.15) can be estimated from above by the sum of the probabilities of these events. The probability of the event in (i) is bounded by $(\theta_L^n)^2$ by the BK-inequality. For (ii), we write A_v for the event $\{C_L^n(v) \cap G_L^n \neq \emptyset\}$.

Then the probability of this event is bounded from above by

$$\sum_v \frac{P(v \text{ is open})}{P(v \text{ is closed})} P(C_L^n \cap G_L^n = \emptyset, v \text{ is a closed neighbour of } C_L^n,$$

$$A_v \square A_v \text{ outside } cl(C_L^n))$$

$$= (e^{\lambda 2^{-nd}} - 1) \sum_v \sum_{(\Sigma, \Sigma^*)} P(C_L^n \cap G_L^n = \emptyset,$$

$$A_v \square A_v \text{ outside } \Sigma^* | E(\Sigma, \Sigma^*)) P(E(\Sigma, \Sigma^*))$$

$$= (e^{\lambda 2^{-nd}} - 1) \sum_v \sum_{(\Sigma, \Sigma^*)} P(C_L^n \cap G_L^n = \emptyset | E(\Sigma, \Sigma^*))$$

$$\times P(A_v \square A_v \text{ outside } \Sigma^* | E(\Sigma, \Sigma^*)) P(E(\Sigma, \Sigma^*))$$

$$\leq (e^{\lambda 2^{-nd}} - 1)\theta_L^n \sum_v \sum_{(\Sigma, \Sigma^*)} P(A_v, C_L^n \cap G_L^n = \emptyset | E(\Sigma, \Sigma^*))$$

$$\times P(E(\Sigma, \Sigma^*))$$

$$\leq 2^{nd}(e^{\lambda 2^{-nd}} - 1)\theta_L^n \sum_\Gamma P(G_L^n = \Gamma)$$

$$\times \sum_v P(v \text{ is pivotal for } A_L^n(\Gamma) \text{ and closed})2^{-nd},$$

and it follows from (6.14) that this expression is equal to the desired bound.

$\square$

STEP 4: The next step in our argument is the limit from the finite discrete model to a continuum model in a finite box. Formally, we have not as yet defined a finite volume RCM, so here are the definitions. The model consists of a Poisson process in the box B_L, and any two points of the point process are connected to each other according to the connection function g_L rather than g. Each point is again labelled with probability γ. The component containing the origin is denoted by W_L and the set of labelled vertices by G_L. Of course, we define $\theta_L(\lambda, \gamma)$ as the probability that $W_L \cap G_L \neq \emptyset$, and $\chi_L(\lambda, \gamma)$ as the expected size of W_L on the event that $W_L \cap G_L = \emptyset$. Here are the required limits:

Lemma 6.3 *The function $\theta_L(\lambda, \gamma)$ is differentiable with respect to both $\lambda > 0$ and $\gamma \in (0, 1)$. Furthermore, we have that*

(1) $\lim_{n \to \infty} \theta_L^n = \theta_L$,

(2) $\lim_{n \to \infty} \dfrac{\partial \theta_L^n}{\partial \gamma} = \dfrac{\partial \theta_L}{\partial \gamma}$,

(3) $\displaystyle\lim_{n \to \infty} \frac{\partial \theta_L^n}{\partial \lambda} = \frac{\partial \theta_L}{\partial \lambda}$,

(4) $\displaystyle\lim_{n \to \infty} 2^{-nd} M_L^n = \int_{B_L} g_L(x)\, dx$.

Proof For (1), we write

$$1 - \theta_L^n = \sum_{k=1}^{\infty} (1 - \gamma)^k P(|C_L^n| = k), \qquad (6.16)$$

and

$$1 - \theta_L = \sum_{k=1}^{\infty} (1 - \gamma)^k P(|W_L| = k). \qquad (6.17)$$

It suffices to prove that $P(|C_L^n| = k) \to P(|C_L| = k)$, for all k. The probability that there are two points x and y of the point process X for which $x - y$ is a point of discontinuity of g_L is zero, by Proposition 6.1. Hence it follows that

$$P(|C_L^n| = k \mid X(B_L) = l) \to P(|C_L| = k \mid X(B_L) = l),$$

for all l, which proves (1). From (6.16) and (6.17) we also get that

$$\frac{\partial \theta_L^n}{\partial \gamma} = \sum_{k=1}^{\infty} k(1 - \gamma)^{k-1} P(|C_L^n| = k),$$

and

$$\frac{\partial \theta_L}{\partial \gamma} = \sum_{k=1}^{\infty} k(1 - \gamma)^{k-1} P(|C_L| = k).$$

Hence, also (2) follows from the previous argument. (Note that the differentiability of θ_L with respect to γ is no problem as it is a power series in $(1 - \gamma)$.)

Next, we want to show that θ_L can be differentiated with respect to λ. The right-hand side of (6.16) is just a finite sum because there are only finitely many vertices in B_L. Hence the derivative of the right-hand side of (6.16) is just the sum of the term-by-term derivatives. Therefore, to show that θ_L is differentiable with respect to λ, together with (3), it suffices to prove that $\sum_{k=1}^{\infty} (1 - \gamma)^k (d/d\lambda) P(|C_L^n| = k)$ converges locally uniformly in λ for $n \to \infty$. For this, we again use Russo's formula. The event $\{|C_L^n| = k\}$ is not increasing, but it can be written as the difference of two increasing events: $P(|C_L^n| = k) = P(|C_L^n| \geq k) - P(|C_L^n| \geq k+1)$. We first compute, according to Russo's formula:

$$\frac{d}{d\lambda} P(|C_L^n| \geq k) = 2^{-nd} \exp(-\lambda 2^{-nd})$$
$$\times E(\text{number of pivotal vertices for } \{|C_L^n| \geq k\}).$$

Let us now pause for a moment to realise what we are trying to do. We are approximating the continuum model by a discrete model. In a continuum model, we can define pivotal points in the obvious way, but the notion of closed pivotal points does not make sense. Hence we are interested in *open* pivotal vertices in the discrete model. The state of a vertex is independent of its pivotality and we can write

$$\frac{d}{d\lambda} P(|C_L^n| \geq k) = \frac{2^{-nd} \exp(-\lambda 2^{-nd})}{1 - \exp(-\lambda 2^{-nd})}$$

$$\times \sum_{l=0}^{\infty} E(\text{number of open pivotal vertices for}$$

$$\{|C_L^n| \geq k\} \mid X(B_L) = l) P(X(B_L) = l).$$

Now we use Proposition 1.2 to conclude that the expected number of open pivotal vertices for $\{|C_L^n| \geq k\}$ given $\{X(B_L) = l\}$ is independent of λ, and, because there are only l points in B_L, trivially bounded from above by l. This quantity is denoted by $f_L(k, l, n)$. Using once more the almost everywhere continuity of g_L (Proposition 6.1), it is obvious that $\lim_{n \to \infty} f_L(k, l, n)$ exists and is equal to the corresponding quantity in the continuous model. Finally, we can now write

$$\lim_{n \to \infty} \sum_{k=1}^{\infty} (1 - \gamma)^k \frac{d}{d\lambda} P(|C_L^n| = k)$$

$$= \lim_{n \to \infty} \exp(-\lambda |B_L|) \frac{2^{-nd} \exp(-\lambda 2^{-nd})}{1 - \exp(-\lambda 2^{-nd})}$$

$$\times \sum_{k=1}^{\infty} (1 - \gamma)^k \sum_{l=0}^{\infty} (f_L(k, l, n) - f_L(k+1, l, n)) \frac{(\ell(B_L)\lambda)^l}{l!}.$$

Using the convergence of $f_L(k, l, n)$ and the fact that $f_L(k, l, n)$ is bounded from above by l, for all k and n, this expression is easily seen to converge locally uniformly in λ, for $n \to \infty$.

It remains to prove (4). This is easy though, as we can write

$$2^{-nd} M_L^n = 2^{-nd} \sum_v g_L(v), \tag{6.18}$$

where the sum is over all non-zero vertices in the box B_L in the n-th approximating lattice model. But the right-hand side of (6.18) is just a Riemann sum which converges to $\int_{B_L} g_L(x)\,dx$. □

STEP 5: In this step, we take the so-called *infinite volume limit*, which means that we let L tend to infinity.

Lemma 6.4 *For $\lambda > 0$ and $\gamma \in (0, 1)$, it is the case that*

(1) $\lim_{L \to \infty} \theta_L = \theta$,

(2) $\lim_{L \to \infty} \dfrac{\partial \theta_L}{\partial \gamma} = \dfrac{\partial \theta}{\partial \gamma}$.

Proof It suffices to prove that

$$\lim_{L \to \infty} P(|W_L| = k) = P(|W| = k). \tag{6.19}$$

First, we decompose the event $\{|W| = k\}$ into $\{|W| = k, W \subset B_n\} \cup \{|W| = k, W \not\subset B_n\}$, and similarly for W_L, where $L > n$. From Proposition 6.2 (and its proof) we easily deduce that

$$P(|W_L| = k, W_L \subset B_n) = \frac{\lambda^{k-1}}{(k-1)!} \int_{B_n \times \cdots \times B_n} g_{L,2}(0, x_1, \ldots, x_{k-1})$$

$$\times e^{-\lambda \int_{B_L} g_{L,1}(y; 0, x_1, \ldots, x_{k-1}) dy} d(x_1, \ldots, x_{k-1}), \tag{6.20}$$

where $g_{L,1}$ and $g_{L,2}$ are the analogues of g_1 and g_2 for the connection function g_L. As a consequence of Proposition 6.1, for $(x_1, \ldots, x_{k-1}) \in B_n \times \cdots \times B_n$, it is the case that

$$g_{L,2}(0, x_1, \ldots, x_{k-1}) \to g_2(0, x_1, \ldots, x_{k-1}),$$

as $L \to \infty$. Furthermore, for fixed y and $x_1, \ldots, x_k$, it follows that $1_{B_L}(y) \cdot g_{L,1}(y; 0, x_1, \ldots, x_{k-1}) \to g_1(y; 0, x_1, \ldots, x_{k-1})$, for $L \to \infty$. In order to use dominated convergence for the integrals in the exponents of (6.8) and (6.20), we write

$$g_{L,1}(y; 0, x_1, \ldots, x_{k-1}) \le \sum_{i=0}^{k-1} g_L(y - x_i),$$

where x_0 is just the origin. Furthermore,

$$\int_{B_L} \sum_{i=0}^{k-1} g_L(y - x_i) \, dy \le \sum_{i=0}^{k-1} \sum_{z \in \mathbb{Z}^d} \int_{B_L} g(y - x_i + 2Lz) \, dy$$

$$= \sum_{i=0}^{k-1} \int_{\mathbb{R}^d} g(y - x_i) \, dy < \infty,$$

according to (6.2). From this it follows that (6.20) converges to (6.8) for $L \to \infty$. Finally, we show that $P(|W_L| = k, W_L \not\subset B_n)$ can be made arbitrarily small uniformly for all L large enough, by taking n fixed but large enough. For $0 < N < M < L$, let $E_L(N, M)$ be the event that in the finite volume model in B_L there is at least one direct connection from a point inside B_N to a point outside B_M. For any non-negative integer-valued random variable Y we have $P(Y \geq 1) \leq EY$, and this implies here that

$$P(E_L(N, M)) \leq \lambda^2 \int_{B_N} \int_{B_L \setminus B_M} g_L(y - x) \, dy \, dx$$

$$\leq \lambda^2 \int_{B_N} \int_{\mathbb{R}^d \setminus B_M} g(y - x) \, dy \, dx,$$

where the last inequality follows from arguments as in the proof of (1). Note that this estimate is uniform in L and hence $P(E_L(N, M))$ is small uniformly in L for M large.

Next let $\epsilon > 0$ and take boxes $B_{n_1} \subset \cdots \subset B_{n_{k-1}}$ such that the following events $A_1, \ldots, A_{k-1}$ all have probability at most ϵ, uniformly in L:

$$A_1 = \{\text{the origin is connected to a point outside } B_{n_1}\},$$

$$A_l = \{\text{there is a point inside } B_{n_{l-1}} \text{ connected to a point outside } B_{n_l}\},$$

for $l = 2, \ldots, k - 1$. Now take $n = n_{k-1}$. If $\{|W_L| = k, W_L \not\subset B_n\}$ occurs then there is a point outside B_n connected to the origin in less than k steps. This means that $\cup_{l=1}^{k-1} A_l$ must occur. However, this has probability at most $(k - 1)\epsilon$ for all L and this proves the lemma. $\square$

STEP 6: Finally we are able to prove the desired result. The reader should note that from now on the argument is completely analytic. Let $\lambda_0 < \lambda_H$ and suppose that $\chi(\lambda_0) = \infty$. We write $f_L(\gamma) = \theta_L(\lambda_0, \gamma)$ and $f(\gamma) = \theta(\lambda_0, \gamma)$. If we first combine the two conclusions in Lemma 6.2 and then take the limit for $n \to \infty$ (Lemma 6.3) we obtain

$$f_L \leq \gamma \frac{df_L}{d\gamma} + (f_L)^2 + (1 - \gamma) \left(\int_{B_L} g_L(x) \, dx \right) \lambda_0 (f_L)^2 \frac{df_L}{d\gamma}.$$

From Lemma 6.4 and Proposition 6.1 we then find, taking the limit for $L \to \infty$,

$$f \leq \gamma \frac{df}{d\gamma} + f^2 + (1 - \gamma) \left(\int_{\mathbb{R}^d} g(x) \, dx \right) \lambda_0 f^2 \frac{df}{d\gamma}. \qquad (6.21)$$

We have $f(\gamma) \to 0$ for $\gamma \to 0$, because of Lemma 6.1(i) and the fact that $\lambda_0 < \lambda_H$. Also,

$$\frac{f(\gamma)}{\gamma} \to \infty \qquad (6.22)$$

for $\gamma \to 0$, where we use Lemma 6.1 again, the mean value theorem and the fact that $\chi(\lambda_0) = \infty$. Now let h be the inverse function of f and substitute

$y = f(\gamma)$ and

$$\left(\frac{dh}{dy}\right) = \left(\frac{df}{d\gamma}\right)^{-1}$$

in (6.21) to obtain

$$\frac{1}{y}\frac{dh}{dy} - \frac{1}{y^2}h \le C(1-h) + \frac{dh}{dy},$$

where C is a constant depending on λ_0. Using (6.22) we see that dh/dy is bounded on an interval $(0, b)$ for some $b > 0$ and hence there is a positive constant β such that

$$\frac{1}{y}\frac{dh}{dy} - \frac{1}{y^2}h \le \beta$$

for $0 < y < b$. Integrating this expression from 0 to x with $x \le b$ yields

$$\left[\frac{1}{y}h(y)\right]_0^x \le \beta x$$

and it follows from this and the fact that $h(y)/y \to 0$ when $y \to 0$ (use (6.21)) that for small enough γ,

$$f(\gamma) \ge C'\gamma^{\frac{1}{2}}, \tag{6.23}$$

where C' is a positive constant depending on λ_0.

We let $n \to \infty$ in Lemma 6.2 and obtain, using Lemma 6.3,

$$\theta_L \le \gamma\frac{\partial\theta_L}{\partial\gamma} + (\theta_L)^2 + \lambda\theta_L\frac{\partial\theta_L}{\partial\lambda}.$$

Rewrite this inequality as

$$0 \le (\theta_L)^{-1}\frac{\partial\theta_L}{\partial\gamma} + \gamma^{-1}\frac{\partial}{\partial\lambda}(\lambda\theta_L - \lambda),$$

which can be integrated over $[\epsilon, \delta] \times [\lambda_0, \lambda_1]$ (where $\lambda_1 < \lambda_H$) to obtain

$$0 \le (\lambda_1 - \lambda_0)\log\left(\frac{\theta_L(\lambda_1, \delta)}{\theta_L(\lambda_0, \epsilon)}\right) + (\lambda_1\theta_L(\lambda_1, \delta) - \lambda_1 + \lambda_0)\log\left(\frac{\delta}{\epsilon}\right). \tag{6.24}$$

Now we use Lemma 6.4 and take the limit for $L \to \infty$, which means that we can remove all subscripts L from (6.24). Divide by $\log(\delta/\epsilon)$, let ϵ go to zero and use (6.23) to find

$$0 \le \tfrac{1}{2}(\lambda_1 - \lambda_0) + \lambda_1\theta(\lambda_1, \delta) - \lambda_1 + \lambda_0.$$

If we take the limit for $\delta \to 0$ here, it follows that $\theta(\lambda_1) = \theta(\lambda_1, 0) > 0$, which is the desired contradiction because $\lambda_1 < \lambda_H$. $\qquad\square$

6.4 Uniqueness

In view of the uniqueness results in Boolean models, we may expect the unbounded component in the Poisson RCM to be unique. This is indeed the case:

Theorem 6.3 *In a Poisson RCM (X, g, λ), there is at most one unbounded component a.s.*

The proof of this result is not very different from the corresponding proof in the Boolean model. It is again the ergodicity of the model which guarantees that the number of unbounded components is an a.s. constant. As before, we first show that this number cannot be any finite number apart from zero or one:

Lemma 6.5 *The number of unbounded components in a Poisson RCM is equal a.s. to either zero, one or infinity.*

Proof To derive a contradiction, let us suppose that this number is a.s. equal to $K \geq 2$. This means that there is a box B_n such that there is a positive probability that B_n intersects all of them. Choose $M \leq \infty$ so that $g(x) = 0$ whenever $|x| > M$, noting that M can take the value infinity. Next we partition B_n into at least K cubic cells G_j in such a way that for two neighbouring cells G_1 and G_2 we have $d(x, y) \leq M$ for all $x \in G_1$ and $y \in G_2$. For any subset $A \subset \mathbb{R}^d$, we write (X_A, g, λ) for the RCM obtained by removing all Poisson points outside A and all connections leading to such points. By taking smaller cells if necessary and a possible renumbering of the cells, we can find K cells $G_1, \ldots, G_K$ such that the following event has positive probability, writing $G = \cup_{i=1}^K G_i$:

$$E := \{(X_{G \cup (B_n)^c}, g, \lambda) \text{ contains exactly } K \text{ unbounded}$$
$$\text{components } C_1, \ldots, C_K \text{ such that } C_i \text{ has exactly one}$$
$$\text{Poisson point } x_i \text{ in } G_i, \text{ for } i = 1, \ldots, K\}.$$

Obviously, the following event F also has positive probability:

$$F := \{\text{each of the cells } G_j \text{ in } B_n \text{ outside } G \text{ contains}$$
$$\text{exactly one Poisson point } x_j\}.$$

Now observe that E and F are independent, because they depend on disjoint regions in space. Hence,

$$P(E \cap F) = P(E)P(F) > 0.$$

But given the event $E \cap F$, we can, with positive probability, connect the points x_i and x_j if the cells G_i and G_j share a face, i.e. if they are neighbours, because of the choice of the cell size. However, after doing that, the resulting configuration contains only one unbounded component and this is the desired contradiction. Note that in case $M = \infty$, there is no need to consider the event F at all. In that case we can directly connect all points $x_1, \ldots, x_K$. $\square$

Proof of Theorem 6.3   It remains to rule out the case of infinitely many unbounded components. This, however, can be done in exactly the same way as in the proof of Theorem 3.6. As such, we do not repeat the argument here.

$\square$

In Chapter 7 we shall discuss a uniqueness result for random-connection models driven by general stationary point processes.

6.5 High density

One of the features of continuum percolation models which they do not share with discrete percolation models is the possibility to consider the model at arbitrary high density. When the density λ of the underlying point process tends to infinity, one expects several things to happen. In the first place, larger λ implies that there are on the average more points per unit volume, so it must be easier for the origin to be contained in an infinite component. This should imply that

$$\lim_{\lambda \to \infty} \theta_g(\lambda) = 1. \tag{6.25}$$

In fact, a proof of (6.25) is not hard. We shall, however, prove a much stronger result below. In addition to the probability of the origin being in an infinite component, one can consider the distribution of finite components. It seems reasonable to guess that the probability for a point to be isolated (i.e. not connected to any other point) given the fact that its component is finite converges to 1 for $\lambda \to \infty$. In other words, 'most' finite components should consist of only one point. We shall prove the following theorem:

Theorem 6.4   *Suppose g satisfies (6.2). Then we have*

$$\lim_{\lambda \to \infty} \frac{-\log(1 - \theta_g(\lambda))}{\lambda \int_{I\!\!R^d} g(x)\, dx} = 1. \tag{6.26}$$

Before giving the proof of this result, let us spend a few words on it. In fact, the assertion of the theorem implies that $1 - \theta_g(\lambda) \sim \exp(-\lambda \int_{I\!\!R^d} g(x)\, dx)$, which is equal to the probability that the origin is isolated. So not only does Theorem 6.4 imply (6.25), it also asserts that the rate at which $\theta(\lambda) = \theta_g(\lambda)$ tends to 1 corresponds to the rate at which the probability of being isolated tends to zero. The proof of Theorem 6.4 is based on the following lemma.

Lemma 6.6 *Suppose that g has bounded support. Then we have*

$$\lim_{\lambda \to \infty} \frac{(1 - \theta_g(\lambda))}{P_\lambda(|W| = 1)} = 1.$$

Before giving the proof of Lemma 6.6, we demonstrate how Theorem 6.4 follows from it. Let g satisfy (6.2), and let $g_r(x) := g(x)1_{\{|x| \leq r\}}$. A simple coupling argument shows that

$$1 - \theta_g(\lambda) \leq 1 - \theta_{g_r}(\lambda).$$

From the fact that $1 - \theta_g(\lambda) \geq P_{(\lambda,g)}(|W| = 1)$ and Lemma 6.6 we thus find

$$1 \leq \liminf_{\lambda \to \infty} \left(\frac{1 - \theta_g(\lambda)}{P_{(\lambda,g)}(|W| = 1)} \right)^{1/\lambda} \leq \limsup_{\lambda \to \infty} \left(\frac{1 - \theta_g(\lambda)}{P_{(\lambda,g)}(|W| = 1)} \right)^{1/\lambda}$$

$$\leq \limsup_{\lambda \to \infty} \left(\frac{1 - \theta_{g_r}(\lambda)}{P_{(\lambda,g)}(|W| = 1)} \right)^{1/\lambda}$$

$$= \limsup_{\lambda \to \infty} \left(\frac{1 - \theta_{g_r}(\lambda)}{P_{(\lambda,g_r)}(|W| = 1)} \cdot \frac{P_{(\lambda,g_r)}(|W| = 1)}{P_{(\lambda,g)}(|W| = 1)} \right)^{1/\lambda}$$

$$= \limsup_{\lambda \to \infty} \left(\frac{\exp(-\lambda \int_{|x| \leq r} g(x)\, dx)}{\exp(-\lambda \int_{\mathbb{R}^d} g(x)\, dx)} \right)^{1/\lambda} = e^{\int_{|x| > r} g(x)\, dx}.$$

Letting $r \to \infty$, we obtain

$$\lim_{\lambda \to \infty} \left(\frac{1 - \theta_g(\lambda)}{P_\lambda(|W| = 1)} \right)^{1/\lambda} = 1.$$

Taking logarithms completes the proof of Theorem 6.4. It remains therefore to prove Lemma 6.6. We write $q_k(\lambda)$ for $P_\lambda(|W| = k)$. In this notation we need to show that

$$\lim_{\lambda \to \infty} \frac{1}{q_1(\lambda)} \sum_{k=1}^{\infty} q_k(\lambda) = 1,$$

or

$$\lim_{\lambda \to \infty} \frac{1}{q_1(\lambda)} \sum_{k=2}^{\infty} q_k(\lambda) = 0.$$

We partition the event that $|W| = k$ as follows. Let $\delta > 0$ and let $(\delta Z)^d$ be the lattice of points of the form δz for $z \in Z^d$. We denote by $F_\delta : \mathbb{R}^d \to \delta Z^d$ the map which sends each point of $\mathbb{R}^d$ to the closest point of $(\delta Z)^d$. This map is well defined for almost all points in $\mathbb{R}^d$. In particular, the image $S = S_\delta = F_\delta(W)$ is well defined with probability 1. We may now write

$$\frac{1}{q_1(\lambda)} \sum_{k=2}^{\infty} q_k(\lambda) = \frac{1}{q_1(\lambda)} \sum_{m=1}^{\infty} \sum_{k=2}^{\infty} P_\lambda(|W| = k, |S_\delta| = m). \tag{6.27}$$

It therefore suffices to prove the three following propositions:

Proposition 6.4 *Suppose g has bounded support. For $\delta > 0$ sufficiently small we have*

$$\lim_{\lambda \to \infty} \frac{1}{q_1(\lambda)} \sum_{k=2}^{\infty} P_\lambda(|W| = k, |S_\delta| = 1) = 0.$$

Proposition 6.5 *Suppose g has bounded support. For $\delta > 0$ sufficiently small, there exists an m_0 such that*

$$\lim_{\lambda \to \infty} \frac{1}{q_1(\lambda)} \sum_{m=m_0}^{\infty} \sum_{k=2}^{\infty} P_\lambda(|W| = k, |S_\delta| = m) = 0.$$

Proposition 6.6 *Suppose g has bounded support. For $\delta > 0$ sufficiently small we have for each fixed m,*

$$\lim_{\lambda \to \infty} \frac{1}{q_1(\lambda)} \sum_{k=2}^{\infty} P_\lambda(|W| = k, |S_\delta| = m) = 0.$$

Of course, Proposition 6.6 is stronger than Proposition 6.4, but the latter will be used in the proof of the former.

Proof of Proposition 6.4 We define $q_k^\delta(\lambda) = P_\lambda(|S_\delta| = 1, |W| = k)$. When no confusion can arise, we drop sub- and superscripts. When $|W| = 1$, $|S|$ can be 1 only if all points of W are concentrated in $B_{\delta/2}$. Hence from Proposition 6.2 we find

$$\frac{q_k^\delta(\lambda)}{q_1(\lambda)} = \frac{\frac{\lambda^{k-1}}{(k-1)!} \int_{B_{\delta/2}} \cdots \int_{B_{\delta/2}} e^{-\lambda \int_{\mathbb{R}^d} g_1(y; 0, x_1, \ldots, x_{k-1}) \, dy} \, dx_1 \cdots dx_{k-1}}{\exp(-\lambda \int_{\mathbb{R}^d} g(y) \, dy)}$$

$$= \frac{\lambda^{k-1}}{(k-1)!} \int_{B_{\delta/2}} \cdots \int_{B_{\delta/2}} \exp\left(-\lambda \int_{\mathbb{R}^d} (g_1(y; 0, x_1, \ldots, x_{k-1}) \right.$$

$$\left. - g(y)) \, dy\right) dx_1 \cdots dx_{k-1}.$$

From the definition of g_1 and g, $g_1(y; 0, x_1, \ldots, x_{k-1}) - g(y)$ equals the probability that if we take a graph with non-random vertices $0, x_1, \ldots, x_{k-1}$ and y, the vertex y is not isolated but not connected directly to 0 (all with connection function g of course). This expression is therefore bounded from below, for all i, by $g(y - x_i)(1 - g(y)) = g(y - x_i) - g(y)g(y - x_i) \geq (g(y - x_i) - g(y))^+$, where $f^+ = \max\{f, 0\}$ denotes the positive part of the function f. It follows from Proposition 6.3 that

$$\int_{\mathbb{R}^d} (g(y - x_i) - g(y))^+ dy = \tfrac{1}{2} \int_{\mathbb{R}^d} |g(y - x_i) - g(y)| \, dy$$

$$\geq c_i |x_i| \text{ for all } x_i \in B_{\delta/2},$$

for a suitable positive constant c_i. We thus find, writing $B_{\delta/2}^i \subset (\mathbb{R}^d)^{k-1}$ for the set $\{x_1, \ldots, x_{k-1} \in (B_{\delta/2})^{k-1} : |x_i| \geq \max_{j \neq i} |x_j|\}$,

$$\frac{q_k^\delta(\lambda)}{q_1(\lambda)} \leq \sum_{i=1}^{k-1} \frac{\lambda^{k-1}}{(k-1)!} \int_{B_{\delta/2}^i} e^{-\lambda c_i |x_i|} d(x_1, \ldots, x_{k-1})$$

$$= \frac{\lambda^{k-1}}{(k-2)!} \int_{B_{\delta/2}^1} e^{-\lambda c_1 |x_1|} d(x_1, \ldots, x_{k-1})$$

$$= \frac{\lambda^{k-1}}{(k-2)!} \int_{B_{\delta/2}} (\pi_d |x_1|^d)^{k-2} e^{-\lambda c_1 |x_1|} dx_1,$$

where π_d is the volume of the unit ball in $\mathbb{R}^d$. Note that the integrand is a function of $|x_1|$ and we can change variables to obtain

$$\frac{q_k^\delta(\lambda)}{q_1(\lambda)} \leq c_2 \int_0^{\delta/(2\sqrt{d})} \frac{\lambda^{k-1}}{(k-2)!} e^{-\lambda c_1 r} r^{d(k-1)-1} \pi_d^{k-2} dr$$

for a positive constant c_2. Thus,

$$\sum_{k=2}^\infty \frac{q_k^\delta(\lambda)}{q_1(\lambda)} \leq c_2 \lambda \int_0^{\delta/(2\sqrt{d})} e^{-\lambda c_1 r} r^{d-1} \sum_{k=2}^\infty \frac{(\lambda \pi_d)^{k-2}}{(k-2)!} (r^d)^{k-2} dr$$

$$= c_2 \lambda \int_0^{\delta/(2\sqrt{d})} r^{d-1} e^{\lambda \pi_d r^d - \lambda c_1 r} dr. \tag{6.28}$$

We can take δ so small that for all $r \leq \delta/(2\sqrt{d})$ we have $\pi_d r^d \leq \tfrac{1}{2} c_1 r$ and thus the right-hand side of (6.28) is bounded from above by

$$c_2 \lambda \int_0^{\delta/(2\sqrt{d})} e^{-\lambda c_1 r/2} r^{d-1} dr \leq c_2(\lambda)^{-(d-1)} \int_0^\infty e^{-c_1 s/2} s^{d-1} ds.$$

The result follows immediately from this. $\qquad\square$

Proof of Proposition 6.5 We give the proof for the two-dimensional case. It will be clear that a similar proof works in any dimension, but the details become very lengthy to write down. We start by choosing R and δ such that

(i) $g(x) = 0$ for all x with $|x| \geq R$,

(ii) $\delta < \frac{1}{2}R$,

(iii) $g(\frac{1}{2}R + 3\delta) \geq \delta$.

Consider the component W of the origin. If we place a ball $S(x, \frac{1}{2}R)$ with radius $\frac{1}{2}R$ around each point x of W, then, using (i) above, the set $\tilde{F}_W := \cup_{x \in W} S(x, \frac{1}{2}R)$ is a connected set. Denote by F_W the union of all squares of the form $[\delta n, \delta(n+1)] \times [\delta m, \delta(m+1)]$ which intersect $\tilde{F}_W$, where n and m are integers. Then F_W is a bounded set whenever $|W| < \infty$. We denote by ∂F_W the exterior boundary of F_W. The boundary $\gamma = \partial F_W$ consists of a number of edges of length δ, and the number of such edges is denoted by $|\gamma|$. We want to estimate the probability that ∂F_W is a particular curve γ. For such a curve γ, let γ_δ be the set of points in the plane which are at a distance at most δ from γ. We denote by int(γ_δ) the set of points which are in the interior of γ but not in γ_δ. Suppose now that $\partial F_W = \gamma$. We claim that $W \cap \gamma_\delta = \emptyset$ a.s. To see this, suppose that there is a point $x \in W$, and a point $y \in \gamma$ such that $|x - y| \leq \delta$. With probability 1, each point in ∂F_W is not in $\tilde{F}_W$ and hence $y \notin \tilde{F}_W$ a.s. But because of (ii), this is impossible and the claim follows.

Let W_γ be the component of the origin obtained from points in int(γ_δ) and all edges between these points. We define the event $E_{k,\gamma} := \{|W_\gamma| = k, \partial F_{W_\gamma} = \gamma\}$, and the event that there is no direct edge between any point in γ_δ and W_γ is denoted by $E'_{k,\gamma}$. Using Proposition 1.3 we have

$$P(E'_{k,\gamma} \mid E_{k,\gamma}) = \exp\left(-\lambda \int_{\gamma_\delta} g_1(y; W_\gamma)\, dy\right). \tag{6.29}$$

Let y be in γ_δ. If $E_{k,\gamma}$ occurs, then each point on the boundary γ of F_{W_γ} must be closer than 2δ to a point of $\tilde{F}_{W_\gamma}$. Hence each point of γ_δ must be closer than 3δ to $\tilde{F}_{W_\gamma}$. Thus, each $y \in \gamma_\delta$ must be closer than $3\delta + \frac{1}{2}R$ to a point in W_γ and it follows from (iii) that $g_1(y; W_\gamma) \geq \delta$. Also, the volume of γ_δ can be estimated using the observation that each edge of γ has a square of side length $\frac{1}{2}\delta$ centered at the midpoint of the edge and which are disjoint for different edges. Thus, the Lebesgue measure $\ell(\gamma_\delta)$ is at least $|\gamma|\delta^2/4$. We now obtain from (6.29):

$$P(E'_{k,\gamma} \mid E_{k,\gamma}) \leq \exp(-\lambda|\gamma|\delta^3/4).$$

It follows that

$$\sum_{k=1}^{\infty} P(|W| = k, \partial F_W = \gamma) \le \exp(-\lambda|\gamma|\delta^3/4) \sum_{k=1}^{\infty} P(E_{k,\gamma})$$

$$\le \exp(-\lambda|\gamma|\delta^3/4). \tag{6.30}$$

The rest of the argument consists of classical counting arguments. We need two facts here: the area enclosed by a curve γ, which in our case is piecewise linear, is at most some constant c_1 (independent of γ) times $|\gamma|^2$, and the number of closed curves γ with $|\gamma| = m$ along edges of the square lattice and which encloses the origin is at most c_2^m for some positive constant c_2; see the proof of Theorem 1.1. We may now write

$$\sum_{k=1}^{\infty} P(|W| = k, |S| \ge m_0) \le \sum_{k=1}^{\infty} P(|W| = k, |\gamma| \ge c_1 m_0^{1/2})$$

$$= \sum_{m \ge c_1 m_0^{1/2}} \sum_{k=1}^{\infty} P(|W| = k, |\gamma| = m)$$

$$= \sum_{m \ge c_1 m_0^{1/2}} \sum_{k=1}^{\infty} \sum_{\gamma:|\gamma|=m} P(|W| = k, \partial F_C = \gamma)$$

$$\le \sum_{m \ge c_1 m_0^{1/2}} \exp(-\lambda m \delta^3/4) c_2^m,$$

using (6.30). Now using the fact that

$$q_1(\lambda) = \exp\left(-\lambda \int_{\mathbb{R}^d} g(x)\, dx\right)$$

$$= \exp(-c_3\lambda), \text{ say}$$

we find

$$\frac{1}{q_1(\lambda)} \sum_{m=m_0}^{\infty} \sum_{k=1}^{\infty} P(|W| = k, |S| \ge m_0)$$

$$\le \sum_{m \ge c_1 m_0^{1/2}} e^{c_3\lambda}(c_2 e^{-\lambda\delta^3/4})^m$$

$$\le e^{c_3\lambda} \frac{(c_2 \exp(-\lambda\delta^3/4))^{c_1 m_0^{1/2}}}{(1 - c_2 \exp(-\lambda\delta^3/4))}.$$

When $\frac{1}{4}\delta^3 c_1 m_0^{1/2} > c_3$, this tends to zero for $\lambda \to \infty$, proving the proposition. $\qquad\qquad\qquad\qquad\qquad\qquad\qquad\qquad\qquad\qquad\qquad\qquad\qquad$ $\square$

Proof of Proposition 6.6 Again, we shall give the proof for the two-dimensional case, as the higher-dimensional case can be proved similarly. From the fact that g has bounded support, it follows that there are only finitely many configurations for S such that $|S| = m$. So it is enough to show that for any finite subset η of $\delta \mathbb{Z}^2$, we have

$$\lim_{\lambda \to \infty} \frac{1}{q_1(\lambda)} \sum_{k=2}^{\infty} P(|W| = k, S = \eta) = 0.$$

We have to introduce some more notation. We denote by W_η the component of the origin after removing all points outside $F_\delta^{-1}(\eta)$ (and the edges leading to such points). We denote by $E_{\eta,k}$ the event that $|W_\eta| = k$ and that $F_\delta(W_\eta) = \eta$. Furthermore, H_η is defined to be the event that no point of the point process X in $\mathbb{R}^2 \backslash F_\delta^{-1}(\eta)$ is connected to any point in W_η. Then,

$$\{|W| = k, S = \eta\} = \{E_{\eta,k} \cap H_\eta\}.$$

We want to estimate the probability of H_η given $E_{\eta,k}$. Let δ_1 be so small that the conclusion of Proposition 6.4 holds for $2\delta_1$. Define r_η as the highest first coordinate of a point in η: $r_\eta = \max\{x_1 : (x_1, x_2) \in \eta\}$. The points l_η and t_η are defined similarly for the lowest first coordinate and highest second coordinate, respectively. The set $A_r \subset \mathbb{R}^2$ is defined as $\{(x_1, x_2) : x_1 > r_\eta + \frac{1}{2}\delta\}$. Furthermore, $A_l := \{(x_1, x_2) : x_1 < l_\eta - \frac{1}{2}\delta\}$ and $A_t := \{(x_1, x_2) : x_2 > t_\eta + \frac{1}{2}\delta\}$. Note that A_r and A_l are disjoint. Given the event $E_{\eta,k}$, the probability that H_η occurs is (using Proposition 1.3)

$$\exp\left(-\lambda \int_{\mathbb{R}^2 \backslash F_\delta^{-1}(\eta)} g_1(y; W_\eta)\, dy\right). \tag{6.31}$$

Let $\delta > 0$ be much smaller than δ_1. If $E_{\eta,k}$ occurs, then there is at least one point $(x_1, x_2) \in W_\eta$ for which $|x_1 - r_\eta| \le \frac{1}{2}\delta$ and we have

$$\int_{A_r} g_1(y; W_\eta)\, dy \ge \int_{(\delta,\infty) \times \mathbb{R}} g(y)\, dy,$$

and similarly,

$$\int_{A_l} g_1(y; W_\eta)\, dy \ge \int_{(-\infty, -\delta) \times \mathbb{R}} g(y)\, dy.$$

So,

$$\int_{A_l \cup A_r} g_1(y; W_\eta)\, dy \ge \int_{\mathbb{R}^2} g(y)\, dy - \int_{(-\delta,\delta) \times \mathbb{R}} g(y)\, dy. \tag{6.32}$$

Let $A_+ = (0, \delta) \times (\delta, \infty)$ and $A_- = (-\delta, 0) \times (\delta, \infty)$ and take δ so small that

$$\int_{A_-} g(y)\, dy = \int_{A_+} g(y)\, dy \geq \int_{(-\delta, \delta) \times \mathbb{R}} g(y)\, dy + c_1$$

for some positive c_1. If both the width $r_\eta - l_\eta$ and the height (defined similarly) of η are smaller than $2\delta_1$, we are done by Proposition 6.4. So without loss of generality, we may assume that the width $r_\eta - l_\eta$ is at least $2\delta_1$. If $E_{\eta,k}$ occurs, then there is at least one point $(x_1, x_2) \in W_\eta$ such that $|t_n - x_2| \leq \frac{1}{2}\delta$. Either $A_+ + (0, x_2)$ or $A_- + (0, x_2)$ is contained in $\mathbb{R}^2 \setminus F_\delta^{-1}(\eta)$ and hence we find, using (6.31) and (6.32),

$$P(H_\eta \mid E_{\eta,k}) \geq \int_{\mathbb{R}^2} g(y)\, dy + c. \tag{6.33}$$

Thus

$$\frac{1}{q_1(\lambda)} \sum_{k=2}^{\infty} P(|W| = k, S = \eta) = e^{-\lambda \int_{\mathbb{R}^d} g(y)\, dy} \sum_{k=2}^{\infty} P(E_{\eta,k} \cap H_\eta)$$

$$\leq \sum_{k=2}^{\infty} P(E_{\eta,k}) e^{-\lambda c_1}.$$

This tends to zero when $\lambda \to \infty$. $\qquad\qquad\qquad\qquad\qquad\qquad\qquad\square$

6.6 Notes

The material in Sections 6.1 and 6.5 is taken from Penrose (1991). Proposition 6.1 is from Meester (1995) and Propositions 6.2 and 6.3 are taken from Penrose (1991). The equality of the critical densities is due to Meester (1995), but also Sarkar (1994) obtained the equality for a restricted class of connection functions. The argument given here is a continuum version of the argument given by Aizenman and Barsky (1987) for discrete percolation. The uniqueness of the unbounded component appears in Burton and Meester (1993).

7

Models driven by general processes

The case in which the driving point process is Poisson has been studied extensively in the previous chapters. It is natural to investigate what happens when the underlying process is not necessarily Poisson. Many of the results obtained for Poisson processes seem to depend heavily on the independence structure of such processes. However, sometimes it turns out that it is stationarity rather than independence which makes an argument work. In such cases, the assumption of independence obscures the picture of what is really happening. The assumption that a point process be stationary is, in fact, very weak. It turns out that many proofs require, in addition to stationarity, that the point process also be ergodic. The class of ergodic point processes is much smaller than the class of stationary ones, so this seems to be a real loss of generality. However, in the first section we shall treat an interesting technique which makes it possible to carry over results for ergodic models to stationary ones. In this chapter, unless we specify that a Boolean model is Poisson or ergodic, it will only be assumed that it is stationary.

7.1 Ergodic decomposition

Consider a measurable space $(\Omega, \mathcal{F})$ and let T be a transformation from Ω into itself. We denote by $\mathcal{M}_T$ the set of all probability measures μ on $(\Omega, \mathcal{F})$ which make $(\Omega, \mathcal{F}, \mu, T)$ into a measure-preserving dynamical system. Our aim here is to show that the ergodic measures in $\mathcal{M}_T$ are very special in the following sense:

Proposition 7.1 *The set $\mathcal{M}_T$ is convex and the ergodic measures (w.r.t. T) are exactly the extremal points of $\mathcal{M}_T$.*

181

Proof For convexity we need to show that whenever μ_1 and μ_2 are in $\mathcal{M}_T$, so is $\alpha\mu_1 + (1-\alpha)\mu_2$ for all $0 \le \alpha \le 1$, where $(\alpha\mu_1 + (1-\alpha)\mu_2)(E) = \alpha\mu_1(E) + (1-\alpha)\mu_2(E)$ for all $E \in \mathcal{F}$. This, however, follows immediately from the definition of measure-preserving transformations.

To prove the second assertion of the proposition, suppose that $\mu \in \mathcal{M}_T$ is not ergodic. This means that the σ-algebra of T-invariant sets is not trivial. Hence there exists a T-invariant measurable set E such that $0 < \mu(E) < 1$. We can define two probability measures μ_1 and μ_2 such that for every $A \in \mathcal{F}$ we have

$$\mu_1(A) = \frac{\mu(A \cap E)}{\mu(E)} \quad \text{and} \quad \mu_2(A) = \frac{\mu(A \cap E^c)}{\mu(E^c)}.$$

Because

$$\mu_1(T^{-1}A) = \frac{\mu(T^{-1}A \cap E)}{\mu(E)} = \frac{\mu(T^{-1}A \cap T^{-1}E)}{\mu(E)}$$
$$= \frac{\mu(T^{-1}(A \cap E))}{\mu(E)} = \frac{\mu(A \cap E)}{\mu(E)} = \mu_1(A),$$

it follows that T is measure-preserving w.r.t. μ_1. (We also say that μ_1 is T-invariant in such a case.) A similar argument is valid for μ_2. Note that $1 = \mu_1(E) \ne \mu_2(E) = 0$ whence $\mu_1 \ne \mu_2$. But we have

$$\mu(A) = \mu(E)\mu_1(A) + (1 - \mu(E))\mu_2(A)$$

and hence μ is not an extremal point of $\mathcal{M}_T$. Conversely, suppose that μ is not extremal but ergodic. Then $\mu = \alpha\mu_1 + (1-\alpha)\mu_2$ for some $0 < \alpha < 1$ and $\mu_1 \ne \mu_2$ both in $\mathcal{M}_T$. This of course implies that μ_1 is absolutely continuous w.r.t. μ, whence the Radon–Nikodym theorem guarantees the existence of a μ-integrable function f such that

$$\mu_1(A) = \int_A f\,d\mu \tag{7.1}$$

for all $A \in \mathcal{F}$. Both μ_1 and μ_2 are T-invariant and it follows easily that f is T-invariant also, i.e. $f(T) = f$ μ-a.s. It is well known (see e.g. Petersen 1983 Proposition 4.1, p. 42) that this implies that f is an a.s. constant, which in turn implies by (7.1) that $f \equiv 1$ a.s. Hence $\mu = \mu_1$ and hence either $\alpha = 1$ or $\mu_1 = \mu_2$, a contradiction in either case. $\square$

Given the structure of a convex set of invariant measures, the extremal points of which are exactly the ergodic ones, it should not come as a surprise that it is possible to 'decompose' any T-invariant measure into ergodic ones. We make this precise in a moment, but first let us return to stationary point processes and start with an elementary example of such a decomposition.

Consider two independent Poisson point processes X_1 and X_2 on $I\!R^d$ with densities λ_1 and λ_2, respectively, where $\lambda_1 \neq \lambda_2$. Let X be the point process defined to be equal to X_1 with probability $\frac{1}{2}$ and equal to X_2 with probability $\frac{1}{2}$. In other words, with probability $\frac{1}{2}$, $X(A) = X_1(A)$ for all $A \subset I\!R^d$ simultaneously, and with probability $\frac{1}{2}$, $X(A) = X_2(A)$ for all $A \subset I\!R^d$. Then X is a stationary point process which is *not* ergodic. To see the latter fact, consider the event $E = \{\lim_{t \to \infty} t^{-d} X([0, t]^d) = \lambda_1\}$. According to Proposition 2.4 and 2.6 and the fact that $\lambda_1 \neq \lambda_2$, the probability of E is $\frac{1}{2}$. It is an easy matter to check that E is translation invariant and hence X cannot be ergodic. In the notation of Chapter 2, if μ is the measure on $(\Omega, \mathcal{F})$ corresponding to X and μ_1 and μ_2 the measures corresponding to X_1 and X_2, respectively, then $(\Omega, \mathcal{F}, \mu, S_t)$ is not an ergodic m.p. dynamical system, but it is the case that $\mu = \frac{1}{2}\mu_1 + \frac{1}{2}\mu_2$. Furthermore $(\Omega, \mathcal{F}, \mu_i, S_t)$, $i = 1, 2$, are ergodic systems (Proposition 2.6).

The following proposition shows that this construction can be carried through in much greater generality than the example above. For a proof of the proposition we refer to Denker, Grillenberger and Sigmund (1976, section 13).

Proposition 7.2 *Let $(\Omega, \mathcal{F}, \mu, T)$ be an m.p. dynamical system and let f be a real, μ-integrable function on Ω. There is a set $E \in \mathcal{F}$ with $\mu(E) = 0$ such that for all $\omega \in \Omega \backslash E$, there exists an ergodic measure μ_ω on $(\Omega, \mathcal{F}, T)$ such that $\omega \to \int_\Omega f d\mu_\omega$ is $\mathcal{F}$-measurable, f is μ_ω-integrable and*

$$\int_\Omega f d\mu = \int_{\Omega \backslash E} \int_\Omega f d\mu_\omega d\mu(\omega). \tag{7.2}$$

The family of measures $(\mu_\omega)_{\omega \in \Omega \backslash E}$ is called the *ergodic decomposition* of μ. (There is a certain uniqueness of the ergodic decomposition, that is why we call it *the* ergodic decomposition. We will not be concerned with this here though.) Note that we do not require that the μ_ω's be different for different values of ω. Indeed, in the example given, $\mu_\omega = \mu_1$ whenever ω is such that $X(A) = X_1(A)$ for all A, and $\mu_\omega = \mu_2$ otherwise. As a special case of (7.2), consider the case where $f = 1_A$ for some $A \in \mathcal{F}$. Then (7.2) reduces to

$$\mu(A) = \int_{\Omega \backslash E} \mu_\omega(A) d\mu(\omega). \tag{7.3}$$

Note that by taking $A = \Omega$ in (7.3), we see that for almost all $\omega \in \Omega \backslash E$ we have

$$\mu_\omega(\Omega) = 1. \tag{7.4}$$

As an application of (7.3), consider a Boolean model $(\Omega, \mathcal{F}, P)$, where $\Omega = \Omega_1 \times \Omega_2$ and $P = P_1 \times P_2$ as in Chapter 1. What form does the ergodic decomposition of this Boolean model have? For this, consider the point process P_1 defined on $(\Omega_1, \mathcal{F}_1)$ (the notation is as in Chapter 2, Section 2.1). As already observed in Chapter 2, we may assume that an ergodic point process is ergodic under translation by e_1. This point process has, according to (7.3), an ergodic decomposition $(P_{1,\omega_1})_{\omega_1 \in \Omega_1 \setminus E}$, where $E \in \mathcal{F}_1$ satisfies $P_1(E) = 0$. From (7.4) it follows that $(\Omega_1, \mathcal{F}_1, P_{1,\omega_1})$ is an (ergodic) point process. From Proposition 2.8 we know that any Boolean model driven by an ergodic point process is ergodic. Hence the Boolean model $(\Omega_1, \mathcal{F}_1, P_{1,\omega_1}) \times (\Omega_2, \mathcal{F}_2, P_2)$ is ergodic. We conclude that $(P_{1,\omega_1} \times P_2)_{\omega_1 \in \Omega_1 \setminus E}$ is the ergodic decomposition of the Boolean model $P_1 \times P_2$. In particular, the distribution of the radii is the same for all ergodic components of the decomposition. A similar remark applies to random-connection models; the connection function is the same in almost all components of the ergodic decomposition.

Returning specifically to Boolean models, suppose we can show that for all *ergodic* Boolean models with a certain property Q, an event A occurs almost surely and that we are faced with the problem of extending this result to stationary Boolean models. Given any stationary Boolean model with property Q, we use (7.3) and conclude that whenever $P_\omega(A) = 1$ for all $\omega \in E$, then also $P(A) = 1$, *provided that almost all elements in the ergodic decomposition of P satisfy property Q.* (We emphasize the latter statement because this is a necessary part of the argument which is sometimes forgotten in the literature.) Ergodic decomposition, therefore, provides a technique to extend results from ergodic models to general stationary models, but care is needed throughout this procedure.

7.2 Basic facts on coverage

Before we investigate percolation properties of general models, we prove some facts which are either interesting in themselves or which will be useful in later sections. The first result is a generalisation of a result which we already proved for Poisson Boolean models in Proposition 3.1.

Proposition 7.3 *Consider a Boolean model (X, ρ) in $\mathbb{R}^d$. If $E\rho^d = \infty$ then the whole space is covered by balls a.s.*

Proof As noticed in the previous section, the ergodic components of (X, ρ) under T_{e_1} say, all have the same radius distribution ρ and we can henceforth assume that the model is ergodic with respect to T_{e_1} (which implies ergodicity under the group of all translations).

If X has infinite density, we can 'thin' the process in some stationary way so as to obtain a finite density process. If we prove the proposition for this process, then it is certainly true for the original infinite density process. Therefore, we may assume that the density $\lambda(X)$ of X is finite and is equal to 1. Let C_n be the ball centred at the origin and with radius $2^{n/d}, n \in N$. (C_n is non-random.) Note that $\ell(C_{n+1}) = c_d 2^{n+1} = 2\ell(C_n)$, where c_d is a constant depending only on the dimension. From Proposition 2.4 we have that for n large enough (depending on the realisation) $\frac{3}{4}V_n \leq X(C_n) \leq \frac{5}{4}V_n$, where $V_n = \ell(C_n)$. Now we write, for large enough n, $X(C_{n+1}\backslash C_n) = X(C_{n+1}) - X(C_n) \geq \frac{3}{4}V_{n+1} - \frac{5}{4}V_n = \frac{6}{4}V_n - \frac{5}{4}V_n = \frac{1}{4}V_n$. Hence, for n large enough the 'annulus' $C_n\backslash C_{n-1}$ contains at least $b_d 2^{n+1}$ points of the point process, where b_d is another constant depending only on the dimension.

Now let E_n be the event that C_0 is *not* completely covered by a ball which is centred in $C_n\backslash C_{n-1}$. Furthermore, let A_m be the event that m is the first index such that $X(C_n\backslash C_{n-1}) \geq b_d 2^{n+1}$ for all $n \geq m$. It follows from the above that up to a set of measure 0, the A_m's form a partition of the probability space. Write

$$P\left(\bigcap_{k=m}^{\infty} E_k \mid A_m\right) \leq P\left(\bigcap_{k=m}^{\infty} \text{(all balls centred in } C_k\backslash C_{k-1} \text{ have}\right.$$

$$\left. \text{radius at most } 2^{k/d} + 1) \mid A_m\right)$$

$$\leq \prod_{k=m}^{\infty} P(\rho \leq 2^{k/d} + 1)^{b_d 2^{k+1}},$$

where the last inequality follows from the independence of the radii and the point process. It suffices to show that this expression equals zero. For k large enough, $2^{k/d} + 1 \leq 2^{(k+1)/d}$, so if we replace $k+1$ by k, for k large, each term in the product is at most $P(\rho \leq 2^{k/d})^{b_d 2^k}$. Now, for m large enough, we find

$$\prod_{k=m}^{\infty} P(\rho \leq 2^{k/d})^{b_d 2^k} = \prod_{k=m}^{\infty} P(\rho^d \leq 2^k)^{b_d 2^k}$$

$$\leq \prod_{k=m}^{\infty} \left\{ P(\rho^d \leq 2^k) \cdot P(\rho^d \leq 2^k + 1) \cdots \right.$$

$$\left. P(\rho^d \leq 2^{k+1} - 1) \right\}^{b_d}$$

$$= \left\{ \prod_{k=2^m}^{\infty} P(\rho^d \le k) \right\}^{b_d}$$

$$= \left\{ \prod_{k=2^m}^{\infty} \left(1 - P(\rho^d > k) \right) \right\}^{b_d}.$$

This expression is zero if and only if $\sum_{k=2^m}^{\infty} P(\rho^d > k) = \infty$, which is equivalent to $E\rho^d = \infty$, proving the proposition. $\square$

The converse of the last proposition is not true in general (but it is in the Poisson case, see Chapter 3). It is not hard to show, using the same argument as in (3.2), that if $E\rho^d < \infty$ and $\lambda(X) < \infty$, then the expected number of balls intersecting a bounded region is finite. Of course the condition that the density of X be finite is necessary: if the density of X is infinite, then the expected number of points in a bounded region is infinite and so is the expected number of balls intersecting this bounded region. However, if we only want the number of intersecting balls to be finite a.s. then we do not need the finite-density assumption:

Proposition 7.4 *Consider an ergodic Boolean model (X, ρ) in $\mathbb{R}^d$ such that the probability that the whole space is completely covered is strictly smaller than 1 (and hence equal to 0 by ergodicity). Then any bounded region in $\mathbb{R}^d$ is intersected by only finitely many balls a.s.*

Proof It suffices to consider the unit circle as the bounded region, so suppose that infinitely many balls have non-empty intersection with the unit circle with positive probability. Look at the proof of Proposition 3.1 and observe that the only properties of the Poisson process we have used there are the fact that the Poisson process is locally finite and the ergodicity of the transformations S_{e_i}. The former is a property shared by all point processes and the latter can be generalised too: it follows from Proposition 2.7 that we can choose an orthonormal base $(e_1, \ldots, e_d)$ of $\mathbb{R}^d$ such that all transformations S_{e_i} act ergodically. (Again, the notation is as in Chapter 2.) Consider random variables Y_n and Z_n, $n \in Z$, defined as in the proof of Proposition 3.1 but now with respect to this new base. It then follows from that proof that the whole space is covered a.s., which is the required contradiction. $\square$

Note that the requirement that the process is ergodic in the last proposition is necessary: the conclusion is false for stationary processes. To see this, take a mixture of two ergodic processes, one where vacancy exists a.s. and one where bounded regions are intersected by infinitely many balls a.s.

7.3 Unbounded components in Boolean models

We consider a Boolean model (X, ρ) where X is a stationary point process in $I\!R^d$ and ρ is the radius random variable of the model. Our aim here is to provide a classification of the possible topological structure of unbounded components in this model. In this section, C denotes a component which can be either occupied or vacant. Often, we consider the complement of a component C and, in particular, we are interested in the connected components of this complement in the usual topological sense. The latter components have nothing to do with the components in the Boolean model; so in order to avoid any confusion, we shall refer to these connected components as connected *sets*. Thus, the complement of a (vacant or unbounded) component of the Boolean model is the union of its connected sets.

Definition 7.1 *Let C be a component of the Boolean model. Then the* interior *of C is defined as*

$$\text{int}\,(C) := \cup\{K;\ K \text{ is a bounded connected set in } I\!R^d\backslash C\}.$$

The closure *is defined as*

$$\text{cl}(C) := C \cup \text{int}\,(C),$$

and the exterior *is defined as*

$$\text{ext}(C) := \cup\{K;\ K \text{ is an unbounded connected set in } I\!R^d\backslash C\}.$$

If for two components C_1 and C_2 we have

$$C_1 \subseteq \text{cl}(C_2),$$

we say that C_1 is *enclosed* by C_2 and we write $C_1 \prec C_2$. The relation '$\prec$' defines a partial ordering on the set of all components.

Lemma 7.1 *The maximal components with respect to the ordering '$\prec$' are exactly all unbounded components a.s.*

Proof It is enough to prove the lemma for all ergodic components of (X, ρ) so we can assume ergodicity of the Boolean model. If $I\!R^d$ is completely covered by balls, we are done, so suppose it is not. From Proposition 7.4 we have that only finitely many balls intersect any bounded area a.s. This implies that any bounded occupied component has strictly positive distance to the nearest other occupied component a.s. In particular, for any bounded occupied component C there is some $\epsilon > 0$ (depending on C) such that the distance from C to the nearest other occupied component is at least 2ϵ, say. (Note that both C and ϵ depend on the configuration.) This means that the set $\{x \in \mathbb{R}^d : 0 < d(x, \mathrm{cl}(C)) < \epsilon\}$ is completely vacant. Moreover, this set is connected and is a subset of a vacant component C'. It follows that $C \prec C'$ and C is not maximal indeed. If C is a bounded vacant component, then the boundary of $\mathrm{cl}(C)$ belongs to one occupied component and hence C is not maximal.

It is clear from the definitions that an unbounded component has to be maximal. □

A realisation is said to be *complete* if for each component C, there exists an unbounded component C' such that $C \prec C'$. A realisation is said to be an *infinite cascade* if each component is bounded. As an example of an infinite cascade, consider a Poisson Boolean model with fixed radii in two dimensions at criticality. It follows from Theorem 4.5 and the last paragraph in the Notes to Chapter 4 that in this situation, no unbounded vacant or occupied components exist a.s.

It is quite possible to construct a realisation which is neither complete nor an infinite cascade. However, we have the following result:

Proposition 7.5 *In any Boolean model, the probability of a realisation which is neither complete nor an infinite cascade is zero.*

Proof If the result is true in all ergodic components of the Boolean model we are done, so we again assume ergodicity. Suppose that with positive probability, there exists an infinite sequence of (bounded) components $C_1 \prec C_2 \prec C_3 \prec \cdots$. It will be enough to show that $\cup_{i=1}^{\infty} \mathrm{cl}(C_i) = \mathbb{R}^2$. For this, take any $x \in \mathbb{R}^d$ and fix some $y \in C_1$. If $x \in C_n$ for some n we are done. Otherwise, we draw the straight line segment l from y to x and we take a box B_n which contains the line segment l. We shall colour l with two colours, red and blue, as follows. First, y is coloured red. We move along l in the direction of x and we change colour as soon as we enter a different component C_i. We move farther along l, changing colour again as soon as we enter yet another component and so on. Note that it is quite possible to enter the same component more than once. If the colour changes infinitely often before we reach x then it must be the case that l goes through infinitely many balls. (Here we use the fact that the balls are convex.) But that would imply that infinitely many balls intersect the box B_n and this has probability zero according to Proposition 7.4. So with probability 1, only finitely many changes of colour are possible. Now m changes of colour means that the line segment l could have gone to at most C_{m+1}. In that case, $x \in \mathrm{cl}(C_{m+1})$. Since x is arbitrary, the proof is complete. □

It is clear from the definition that being complete is a translation-invariant property. So if the Boolean model is ergodic, then either almost all realisations are complete or almost all realisations are an infinite cascade. From now on we assume that all Boolean models under consideration are complete.

Definition 7.2 *An N-*branch *of the occupied component W of the origin is a maximal unbounded connected subset of* $W \cap (B_N)^c$.

Before we state the main results of this section we take a closer look at the realisations of Boolean models in two dimensions. Consider three disjoint boxes $B_N^{z_i} (:= B_N + z_i)$, $i = 1, 2, 3$, where $z_i \in 2N\mathbf{Z}^2$. Suppose that in a realisation of (X, ρ), for $i = 1, 2, 3$, the vacant component V_i containing z_i is unbounded and that $\text{ext}(V_i)$ consists of (at least) three unbounded connected sets which we denote by C_j^i, $j = 1, 2, 3$. By taking N larger if necessary (and assuming that the boxes are so far apart that this increase in size causes no intersection among them) we can find points $r_j^i \in C_j^i \cap B_N^{z_i}$ for all i and j and continuous polygonal curves γ_j^i from z_i to r_j^i such that the curve γ_j^i is contained (except its end point r_j^i) in $V_i \cap B_N^{z_i}$ and such that $\gamma_j^i \cap \gamma_{j'}^i = z_i$ whenever $j \neq j'$. We then have the following result, needed later on.

Lemma 7.2 *In a two-dimensional Boolean model it is the case that, in the situation just described, the set*

$$S := \bigcap_{i=1}^{3} \{C_1^i, C_2^i, C_3^i\}$$

has cardinality at most 1.

Proof The proof proceeds by contradiction, using the well-known result on planar graphs by Kuratowski. So suppose that S contains at least two elements C and C', say. This means that we can find j_1, j_2 and j_3 such that $C = C_{j_i}^i$ for $i = 1, 2, 3$ and k_1, k_2 and k_3 such that $C' = C_{k_i}^i$ for $i = 1, 2, 3$. This implies that we can find continuous polygonal curves γ_1 from $r_{j_1}^1$ to $r_{j_2}^2$ and γ_2 from $r_{j_2}^2$ to $r_{j_3}^3$ which are completely contained in C. Let $s_{j_2}^2$ denote the last point (starting from $r_{j_2}^2$) these curves have in common. Similarly, we can find continuous polygonal curves γ_1' from $r_{k_1}^1$ to $r_{k_2}^2$ and γ_2' from $r_{k_2}^2$ to $r_{k_3}^3$ which are completely contained in C'. Let $s_{k_2}^2$ denote the last point (starting from $r_{j_2}^2$) these curves have in common. For ease of notation, we denote by $r_{l_1}^1$ the point in $\{r_1^1, r_2^1, r_3^1\} \backslash \{r_{j_1}^1, r_{k_1}^1\}$. The points $r_{l_2}^2$ and $r_{l_3}^3$ are defined in a similar fashion. Next we consider the bipartite graph G defined as follows. The vertices of G

are $\{z_1, z_2, z_3\} \cup \{s_{j_2}^2, s_{k_2}^2, r_{l_2}^2\}$. The edges are imbedded in the plane along the curves constructed above as follows:

   (i) z_1 is connected to $s_{j_1}^2$ along $\gamma_{j_1}^1$ and γ_1,

   (ii) z_1 is connected to $s_{k_2}^2$ along $\gamma_{k_1}^1$ and γ_1',

   (iii) z_2 is connected to $s_{j_2}^2, s_{k_2}^2$ and $r_{l_2}^2$ along $\gamma_{j_2}^2, \gamma_{k_2}^2$ and $\gamma_{l_2}^2$, respectively,

   (iv) z_3 is connected to $s_{j_2}^2$ along $\gamma_{j_3}^3$ and γ_2,

   (v) z_3 is connected to $s_{k_2}^2$ along $\gamma_{k_3}^3$ and γ_2'.

The construction of the edges of G is such that two edges can intersect only at vertices of G. Now suppose that it is possible to connect $r_{l_2}^2$ with $r_{l_1}^1$ and $r_{l_3}^3$ by arbitrary disjoint curves α and β which do not intersect any of the previously constructed curves (apart from the point $r_{l_2}^2$ of course). This would imply that z_1 and z_3 are connected to $r_{l_2}^2$ along $\gamma_{l_1}^1$ and α, respectively, $\gamma_{l_3}^3$ and β. Thus this would create a new graph G' which is isomorphic to the complete bipartite graph $K_{3,3}$. It is a well-known result in graph theory, due to Kuratowski, that $K_{3,3}$ cannot be imbedded in the plane without non-trivial intersections between the edges. Hence curves α and β do not exist. But this implies that not all three points $r_{l_1}^1, r_{l_2}^2$ and $r_{l_3}^3$ are contained in the exterior of the Jordan curve J connecting $(z_1, r_{j_1}^1, s_{j_2}^2, r_{j_3}^3, z_3, r_{k_3}^3, s_{k_2}^2, r_{k_1}^1, z_1)$ along edges of G. Thus at least one of these three points, $r_{l_1}^1$ say, is contained in the interior of J. However, $r_{l_1}^1$ is contained in the unbounded occupied component $C_{l_1}^1$ and hence there exists a continuous polygonal curve from $r_{l_1}^1$ to infinity which is completely contained in $C_{l_1}^1$. This curve then has to intersect an edge of G, but this is impossible because all edges of G are contained in vacant components or occupied components disjoint from $C_{l_1}^1$. $\square$

    Here are our main results. The first (occupancy) result is true in any dimension, but for the corresponding result for vacancy we need the dimension to be 2.

Theorem 7.1 *Let W denote the occupied component containing the origin. Then for any $N > 0$, the probability that W has more than two N-branches is equal to zero.*

Theorem 7.2 *Consider a Boolean model in two dimensions and let V be the vacant component which contains the origin. The probability that $\mathrm{ext}(V)$ consists of more than two connected sets is equal to zero.*

Before we prove these results, let us look at the consequences in two dimensions. In two dimensions, the fact that any unbounded occupied component C has at most two N-branches implies that ext(C) consists of at most two connected sets. So in two dimensions the exterior of any unbounded component, vacant or occupied, contains at most two connected sets. Each of these connected sets gives rise to one 'neighbouring' unbounded component of the opposite type, and we conclude that each unbounded component has at most two 'neighbouring' unbounded components. With the convention that $\infty - \infty = 0$, this immediately yields:

Theorem 7.3   *In a Boolean model in two dimensions, the number of unbounded occupied components and the number of unbounded vacant components differ by at most one a.s.*

In the last section of this chapter, we shall construct, for any integer K, a Boolean model with K unbounded occupied and $K \pm 1$ unbounded vacant components. The proof of Theorem 7.1 resembles the proof of uniqueness of unbounded components in Poisson Boolean models. In that proof we already used the idea of branches. In the next section we shall see that Theorem 7.3 can be used to prove uniqueness results for Boolean models driven by general point processes under certain conditions on the radius random variable.

Trying to use the ideas of Chapter 3 creates problems similar to those in Chapter 4. We do not want to assume that the density of the point process is finite, and this is crucial in the proof of Theorem 3.6. In the proof of Theorem 4.6, the way out was to look at vacant components inside little cubes. Here, we can do something similar:

Lemma 7.3   *Consider a Boolean model (X, ρ, λ) and choose $K < \infty$ such that $P(\rho > K) > 0$. Let $A \subset \mathbb{R}^d$ be a convex set with diameter at most K. Let $C[A]$ denote the (random) region*

$$C[A] := \bigcup_{x_i \in A} S(x_i, \rho_i);$$

that is, $C[A]$ is the occupied region formed by points in A. Then the number of connected components in $A \cap C[A]$, to be denoted by Y_A, has finite expectation.

Proof   The set A is convex, and hence its intersection with any ball, if not empty, consists of one component exactly. Hence Y_A is at most k if $X(A) = k$. This would complete the proof if the density of X were finite. To treat the general case, note that Y_A can be larger than 1 only if all balls centred in A have radius at most K. Now condition on the number of points in A to obtain

$$E\{Y_A \mid X(A) = k\} = \sum_{n=1}^{\infty} n P(Y_A = n \mid X(A) = k)$$

$$\leq P(Y_A = 1 \mid X(A) = k) + \sum_{n=2}^{k} n\{P(\rho \leq K)\}^k$$

$$\leq 1 + \tfrac{1}{2}k(k+1)\{P(\rho \leq K)\}^k.$$

Hence

$$EY_A \leq \sum_{k=0}^{\infty} \left(1 + \tfrac{1}{2}k(k+1)\{P(\rho \leq K)\}^k\right) P(X(A) = k), \qquad (7.5)$$

and this sum is finite because $P(\rho \leq K) < 1$ by assumption. $\qquad \square$

Proof of Theorem 7.1   It is clear from the discussion on ergodic decomposition in Section 7.1 that it suffices to prove the theorem for ergodic Boolean models. Suppose that W has more than two N-branches with positive probability for some N. The event that W has more than two N-branches is denoted by $E^0(N)$. Choose $\epsilon > 0$ so small that $P(\rho > 2\epsilon\sqrt{d}) > 0$. A *local component* is a connected set in $C[B_\epsilon^{2\epsilon z}] = \cup_{x_i \in B_\epsilon^{2\epsilon z}}(S(x_i, \rho_i) \cap B_\epsilon^{2\epsilon z})$, for some $z \in \mathbf{Z}^d$. From Lemma 7.3, it follows that the expected number of local components in B_ϵ is finite. As in Chapter 3, we choose a number K large. Given K, we choose $M > N$ and define the event $E^0(N, M) := E^0(N) \cap \{$all N-branches of W contain balls centred in at least K different boxes $B_\epsilon^{2\epsilon z} \subset B_{MN}^0 \backslash B_N^0\} \cap \{B_\epsilon$ is covered by a ball centred in $B_\epsilon\}$. It is clear that this event has positive probability η, say, if ϵ is small enough. The event $E^z(N, M)$ is defined by translating this event over the vector z.

It follows from Proposition 2.7 that we can choose ϵ such that all translations $\tilde{T}_{2\epsilon e_i}$ are ergodic. (It might be necessary to rotate the coordinate axes for this.) As in the proof of Theorem 3.6, it follows that the expected cardinality of the set

$$R = \left\{z \in \mathbf{Z}^d \,:\, B_{MN}^{2Nz} \subset B_{LN}^0, \, E^{2Nz}(N, M) \text{ occurs}\right\}$$

is equal to ηL^d. For $z \in R$, we have $Y_{B_\epsilon^{2Nz}} = 1$ by definition. Furthermore, for $z \in R$, we denote by $C_z^{(1)}$, $C_z^{(2)}$ and $C_z^{(3)}$ the set of all local components in each of the first three N-branches (which can be thought of as being ordered in some arbitrary way) contained in $B_{MN}^{2Nz} \backslash B_N^{2Nz}$, then $C_z^{(i)} \cap C_z^{(j)} = \emptyset$ for $i \neq j$, and card($C_z^{(i)}$) $\geq K$ for all i. Finally, for $z \in R$, we identify z with the only local component of B_ϵ^{2Nz}. The rest of the proof now proceeds exactly as in the proof of Theorem 3.6, using local components instead of points of the point process. $\qquad \square$

REMARK: The reader should note that the only place in the proof of Theorem 7.1 where we used the independence of the radii and the point process was to derive the fact that the expected number of local components in B_ϵ is finite. The fact that we put balls around each point rather than some other shape is only relevant as far as this affects the expected number of local components. This means that if we consider any stationary point process and put shapes around each point in a stationary way such that the expected number of local components is finite, the proof above goes through without difficulty.

Proof of Theorem 7.2 In order to arrive at a contradiction, suppose that ext(V) consists of at least three (unbounded) connected sets with positive probability. By looking at the boundary of these connected sets we see that there are at least three unbounded occupied components with positive probability. Hence there is a non-random box B_N such that B_N intersects these three unbounded occupied components with positive probability. We now perform the following trick: consider the two-dimensional lattice $2N\mathbf{Z}^2 = \{2Nz : z \in \mathbf{Z}^2\}$. As in Chapter 1, we can create a stationary point process Y by 'shifting' this lattice over a random vector which is chosen uniformly on B_N. Now consider the superposition of X and Y, where X and Y are chosen to be independent. It is easy to check that the superposition of two independent stationary point processes is again stationary. The points which come from the process Y will not be the centre of a ball with random radius. Instead we do the following: for every point y of the point process Y we check whether or not the square B_N^y has at least three occupied N-branches in the model (X, ρ); i.e. we check whether or not the complement of B_N^y contains at least three unbounded occupied components which intersect the boundary of B_N^y. If this is not the case, we do not put any shape around y. If this is the case, however, then with probability $\frac{1}{2}$, we centre a *square* with side length $2N - \delta$ at y where δ is a small positive number, and with probability $\frac{1}{2}$ we do nothing. The model obtained is stationary, even though the configuration of squares depends on the realisation of the original model (X, ρ).

We do not change the balls which are centred at the points of X. As we already saw above, for $\epsilon > 0$ small enough, the expected number of local components in B_ϵ in the Boolean model (X, ρ) is finite. If, in addition,

$2\epsilon < 2N - \delta$, then the extra squares coming from points of the process Y can increase the number of local components in B_ϵ by at most one, because once a point of Y is centred in B_ϵ, the whole box B_ϵ is covered by the square centred at this point. We conclude from the remark preceding this proof that the new superposed model has all the properties which make the proof of Theorem 7.1 work. Thus, in this superposed model, for any N, there cannot be more than two N-branches in the occupied component of the origin a.s. However, we already concluded that for the Boolean model (X, ρ) there exists a number N such that each box B_N^y intersects three unbounded occupied components with positive probability. There is a positive probability that y is the centre of a square with side length $2N - \delta$. For δ small enough, there is a positive probability that this square intersects all three unbounded occupied components and that the square is contained in B_N. According to Lemma 7.2, for N large enough, there are only finitely many boxes B_N^y, $y \in Y$ which have a non-empty intersection with more than one of the three unbounded occupied components of (X, ρ) which intersect B_N and whose complement contains at least three N-branches. With positive probability, no point of Y in one of these boxes is the centre of a square and hence with positive probability, these unbounded components remain disjoint outside B_N. The conclusion is that we have constructed in the superposed model an unbounded occupied component with three N-branches. According to Theorem 7.1 and the remark preceding this proof, this is impossible and the proof is complete. $\square$

7.4 Uniqueness in Boolean models

In Chapters 3 and 4 we proved that in a Poisson Boolean model both the unbounded occupied and unbounded vacant component are unique a.s. We cannot expect this to be true in general Boolean models. Here is a very simple counterexample in two dimensions. Consider the following point process in the spirit of Example 1 in Section 1.2: we translate the points of the lattice $\{(z_1, 2z_2) : z_1, z_2 \in \mathbf{Z}\}$ over a random (uniform) vector in $[0, 1] \times [1, 2]$. Thus the horizontal distance between two points is 1, and the vertical distance between two points is 2. Let ρ be such that $P(\rho = \frac{2}{3}) = 1$. In this Boolean model it is easy to see that we get infinitely many unbounded occupied and vacant components a.s. Hence certain conditions are necessary in order to obtain uniqueness. The sufficient conditions which we obtain are conditions concerning the support of the radius random variable ρ and moment conditions on the point process X.

Theorem 7.4 *Consider a Boolean model (X, ρ) in $\mathbb{R}^d$. If for every $M > 0$
we have*

$$P(\rho > M) > 0 \tag{7.6}$$

then there is at most one unbounded occupied component a.s.

The situation for vacant components is not quite as nice as this. In two
dimensions however, we have the following result.

Theorem 7.5 *Consider a Boolean model (X, ρ) in two dimensions. If for every
$\epsilon > 0$ we have*

$$P(\rho < \epsilon) > 0 \tag{7.7}$$

then there is at most one unbounded vacant component a.s.

To obtain uniqueness for vacancy in higher dimensions, we have to impose
further conditions on the point process X. But it is easy to rule out all possibilities
except zero, one and infinity:

Theorem 7.6 *For an ergodic Boolean model (X, ρ) in any dimension, if (7.7)
holds, the number of unbounded vacant components is either zero, one or infinity
a.s.*

The last result of this section gives a condition to rule out the case of infinitely
many unbounded vacant components.

Theorem 7.7 *Consider a Boolean model in dimension $d \geq 3$. Suppose that
(7.7) holds and that, in addition,*

$$E((X(B_n))^d) < \infty \tag{7.8}$$

*for all n and that the support of ρ is bounded. Then there is at most one
unbounded vacant component a.s.*

Proof of Theorem 7.4 First we remark that it suffices to prove the result for
ergodic Boolean models. This is clear from the discussion of ergodic decomposition in Section 7.1 and the fact that the distribution of the radii is the same
in almost all ergodic components; see the discussion just before the end of
Section 7.1.

In an ergodic Boolean model, the number of unbounded occupied components is an a.s. constant according to Theorem 2.1. We have to show
that this constant is either zero or one. To this end, suppose first that it is

at least three (which covers the case of infinitely many). This implies that there is a box B_n such that the following event E has positive probability: $E :=$ {there are at least three unbounded occupied components intersecting B_n and $X(B_n) \geq 1$}. A contradiction arises as follows. We choose a point of X in B_n and we increase the radius of the ball centred at this point until this ball intersects at least three unbounded occupied components. Leaving the rest of the realisation as it is, we create in this way a realisation with an unbounded occupied component with three M-branches for some large M. Also, this has positive probability because of the independence of the radii and the point process and the fact that the support of the radii is unbounded. This then contradicts Theorem 7.1.

Next we suppose that the number of unbounded occupied components is finite but larger than one. (Note that there is overlap with the previous case.) Then there exists a number $K \geq 2$ such that the number of unbounded occupied components is equal to K a.s. The argument is almost the same as above: there exists a box B_m such that this box intersects at least two unbounded occupied components and at the same time contains at least one point of X with positive probability. By increasing the radius of a ball centred in this box, we can connect two different unbounded components. In the resulting realisation, we have strictly less than K unbounded occupied components, and all this occurs with positive probability. This contradicts the fact that the number of unbounded occupied components is K a.s. $\qquad\square$

Proof of Theorem 7.5 Again, by the same reason as above, it suffices to prove the theorem for ergodic Boolean models. The proof is similar to the proof of Theorem 7.4, though a little more care is needed. Suppose first that the number of unbounded vacant components is at least three. Then there exist boxes B_n and B_N with $n < N$ such that with positive probability B_n intersects at least three unbounded vacant components and at the same time, all balls centred in B_n are contained in B_N. The fact that the balls can be arbitrarily small now allows us to reduce the radii of all balls centred in B_n until the intersection between any two such balls is empty. All unbounded components in the original realisation which intersected B_n are now connected to each other through the vacancy in B_n, but the configuration outside B_N remains unchanged. Thus no new unbounded vacant components can arise by this procedure. But now we have created an unbounded vacant component whose exterior consists of at least three unbounded connected sets, and this contradicts Theorem 7.2. The case in which the number of unbounded vacant components is finite but larger than one is treated similarly and we omit the proof. $\qquad\square$

Proof of Theorem 7.6   In fact the proof of this theorem has been given already in the proof of Theorem 7.5. It is shown in that proof that we may connect any finite number of unbounded vacant components by reducing the radii of certain balls. So if we first assume that the number of unbounded vacant components (an almost sure constant by ergodicity) is K, say, then by connecting them we see that there is also a strict positive probability of having strictly less than K such components, a contradiction.   □

It remains to prove Theorem 7.7.

Proof of Theorem 7.7   It follows from the ergodic decomposition in (7.2) that (in the notation of (7.2)) if a random variable Y has finite expectation $\int_\Omega Y d\mu$, then almost all expectations $\int_{\Omega_\omega} Y d\mu_\omega$ are also finite. As before, the radius distribution is the same in almost all ergodic components of the Boolean model. Hence we can again assume that the model is ergodic.

We see from (7.8), the boundedness of the balls and Lemma 4.5 that the expected number of vacant components in the unit cube is finite. Check the proof of Theorem 7.1 and observe that if we redefine a local component as a connected vacant component in a cube, the proof goes through completely. Thus we have shown that if V denotes the vacant component of the origin, for any $N > 0$, the probability that V has more than two N-branches is zero.

The rest of the argument is as in Theorem 7.5. According to Theorem 7.6 we only need to rule out the possibility to have infinitely many unbounded vacant components. Indeed, if more than three unbounded vacant components exist with positive probability, we can connect them by reducing the radii of certain balls, thereby obtaining an unbounded vacant component with three N-branches for some N, which is a contradiction.   □

7.5 Uniqueness in random-connection models

We already proved in Chapter 6 that in a Poisson random-connection model, only one unbounded component can exist a.s. The fact that the driving point process is Poisson was used in two places only. First of all, we used the fact that one can 'add' points to an existing configuration. However, we already noted at the end of the proof of Lemma 6.5 that in case the connection function g has infinite range (i.e. satisfies $M = \infty$ in the notation of Chapter 6) we do not need to add these points at all.

The second fact about Poisson point processes we used was that it has finite density. So without any work, we conclude that if the RCM is such that X has finite density and g has infinite range, there can be at most one unbounded component. The assumption for g to have infinite range cannot be omitted. It is easy to construct an RCM with finite-range connection function for which there are infinitely many unbounded components. In fact, the example at the beginning of the previous section can serve here as well, noting that a Boolean model with fixed-radius balls is in fact an RCM too. But we shall see now that it is possible to remove the assumption that the density of X be finite:

Theorem 7.8 *Let (X, g) be an RCM such that g has infinite range. Then there can be at most one unbounded component a.s.*

As noted before, the proof of Lemma 6.5 goes through without difficulty if g has infinite range. Hence we need only to rule out the case of having infinitely many unbounded components. Here we follow the usual strategy and we first deal with the analogue of Lemma 7.3:

Lemma 7.4 *Consider an RCM (X, g) and denote by G the (a.s. finite) graph obtained from this RCM by taking all points of X in B_n and all connections between them. Then the expected number of connected components in G is bounded from above by $(g(2n\sqrt{d}))^{-1} < \infty$.*

Proof First we condition on the event that $X(B_n) = k$ and we shall obtain a bound which is independent of k. We denote the points of X in B_n by $x_1, \ldots, x_k$. Let for $i = 1, \ldots, k$ G_i be the graph which we obtain by only taking into account the points $x_1, \ldots, x_i$ and the connections between these points, so that $G_k = G$. Denote by C_i the number of connected components in G_i. We obviously have that $EC_1 = 1$. To estimate EC_{n+1}, we note that adding a vertex x_{n+1} to $\{x_1, \ldots, x_n\}$ and possible connections between x_{n+1} and $\{x_1, \ldots, x_n\}$ only can increase the number of connected components if x_{n+1} is not connected to any of the previous points. Hence,

$$EC_{n+1} \leq \prod_{i=1}^{n} (1 - g(x_{n+1} - x_i)) \, (EC_n + 1)$$

$$+ \left(1 - \prod_{i=1}^{n} (1 - g(x_{n+1} - x_i))\right) EC_n$$

$$= EC_n + \prod_{i=1}^{n} (1 - g(x_{n+1} - x_i))$$

$$\leq EC_n + (1 - p)^n,$$

where $p := (g(2n\sqrt{d}))^{-1}$. Hence it follows that

$$EC_{n+1} \le \sum_{i=0}^{n} (1 - p)^i \le p^{-1}$$

for all n. This bound is independent of k and the proof is complete. □

Proof of Theorem 7.8   Using the argument in the proof of Lemma 6.5 in Chapter 6 we see that if infinitely many unbounded components exist then, for some N, the component of the origin has with positive probability more than two (disjoint) N-branches, where a branch is defined as in Definition 7.2. If we redefine a local component as a component of the graph obtained by considering the RCM in a box B_N^{2Nz} only (as in Lemma 7.4), we can now copy the proof of Theorem 7.1 to show that this leads to a contradiction. This completes the proof. □

7.6 Cutting and stacking

Stationary and ergodic point processes were introduced in Chapter 1. Some examples were given, but no mechanism was provided to obtain stationary and ergodic point processes in a constructive way. In the examples given in the next section we need a certain type of point processes and the goal of this section is to introduce these point processes. It is quite possible to describe the construction of these processes at a heuristic level. To prove, however, that this construction gives rise to stationary and ergodic point processes, more rigour is needed.

We explain the construction in detail for the two-dimensional case, the higher-dimensional case being a straightforward generalisation of this. The probability space involved is the cube $\Omega := [0, 1)^3$ with ordinary Lebesgue measure which we shall denote by P here. An element $\omega \in \Omega$ is denoted by $\omega = (u, x)$, where $u \in [0, 1)^2$ and $x \in [0, 1)$. The method is such that any suitable (to be made precise later) subset A of Ω gives rise to a stationary and ergodic point process X_A defined on Ω.

Suppose that (u, x) and $(u + u', x)$ are both in Ω. Then we set

$$S_{u'}(u, x) = (u + u', x). \tag{7.9}$$

Thus we have defined the transformation $S_{u'}$ on a subset of Ω. Next, we want to extend this definition to a larger subset of Ω. To this end, we perform the so-called cutting and stacking procedure. We subdivide Ω into four slices

$$A_1^k := [0, 1)^2 \times \left[\frac{k}{4}, \frac{k+1}{4} \right), \quad k = 0, 1, 2, 3.$$

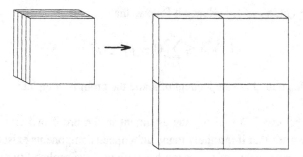

Figure 7.1. The cutting and stacking procedure.

We now rearrange these slices as follows. Let Ω_1 be the set $[0, 2)^2 \times [0, \frac{1}{4})$ and define the 'stacking function' $f_1 : \Omega \to \Omega_1$ as follows

$$f_1(a, b, x) = \begin{cases} (a, 1 + b, x) & \text{for } (a, b, x) \in A_1^0, \\ (1 + a, 1 + b, x - \frac{1}{4}) & \text{for } (a, b, x) \in A_1^1, \\ (a, b, x - \frac{1}{2}) & \text{for } (a, b, x) \in A_1^2, \\ (1 + a, b, x - \frac{3}{4}) & \text{for } (a, b, x) \in A_1^3. \end{cases}$$

Thus the function f_1 just stacks the slice A_1^k to $f_1(A_1^k)$ which is one of the four blocks making up Ω_1; see Figure 7.1. Now we can define $S_{u'}$ on a larger subset of Ω as follows: if both (u, x) and $(u + u', x)$ are in Ω_1 we set

$$S_{u'}(f_1^{-1}(u, x)) = f_1^{-1}(u + u', x). \tag{7.10}$$

Note that this is really an extension of $S_{u'}$. If $S_{u'}$ was already defined in (7.9) it coincides with (7.10).

This procedure can be repeated. We divide Ω_1 into four slices

$$A_2^k := [0, 2)^2 \times \left[\frac{k}{16}, \frac{k+1}{16}\right), \quad k = 0, 1, 2, 3.$$

Putting $\Omega_2 := [0, 4)^2 \times [0, \frac{1}{16})$ we define the stacking function $f_2 : \Omega_1 \to \Omega_2$ as the function which stacks the slice A_2^k to $f_2(A_2^k)$, where $f_2(A_2^k)$ is one of the four blocks making up Ω_2, so

$$f_2(A_2^0) = [0, 2) \times [2, 4) \times [0, \tfrac{1}{16})$$
$$f_2(A_2^1) = [2, 4) \times [2, 4) \times [0, \tfrac{1}{16})$$
$$f_2(A_2^2) = [0, 2) \times [0, 2) \times [0, \tfrac{1}{16})$$
$$f_2(A_2^3) = [2, 4) \times [0, 2) \times [0, \tfrac{1}{16}).$$

If (u, x) and $(u + u', x)$ are both in Ω_2 we set

$$S_{u'}((f_2 \circ f_1)^{-1}(u, x)) = (f_2 \circ f_1)^{-1}(u + u', x). \qquad (7.11)$$

We continue in the obvious way, obtaining sets Ω_n and maps $f_n : \Omega_{n-1} \to \Omega_n$ for all $n \geq 1$ (where $\Omega_0 = \Omega$). At each step we extend the definition of $S_{u'}$. We claim that the subset of points (u, x) of Ω for which $S_{u'}$ is defined for all $u' \in I\!R^2$ has Lebesgue measure 1. To see this, let E_M be the set of points (u, x) for which $S_{u'}$ is defined for all $u' \in I\!R^2$ with $|u'| \leq M$. Furthermore, we put

$$E_M^n := \{\omega \in \Omega : d((f_n \circ f_{n-1} \circ \cdots \circ f_1)(\omega), \partial(\Omega_n)) \geq M\},$$

where $\partial(\cdot)$ denotes the boundary of a set. Then it is easy to see that $E_M^n \subset E_M^{n+1}$ and that $E_M = \cup_{n=1}^\infty E_M^n$. But $P(E_M^n) = (1/2^{2n})(2^n - 2M)^2 \to 1$ as $n \to \infty$. Hence $P(E_M) = 1$ for all M and the claim follows.

Now we can define the point process X_A. For any measurable $B \subset I\!R^2$ consider the set

$$U_B(\omega) := \{u \in B : S_u(\omega) \in A\}. \qquad (7.12)$$

If A is such that $U_B(\omega)$ is almost surely finite for every measurable and bounded set B then we define X_A by the relation

$$X_A(B)(\omega) = \text{card}(U_B(\omega)); \qquad (7.13)$$

i.e. there is a point at x if and only if $S_x(\omega) \in A$.

Proposition 7.6 *Let $A \subset \Omega$ be such that $X_A(B)(\omega)$ is a.s. finite for all bounded sets $B \subset I\!R^2$. Then X_A is a stationary and ergodic point process.*

Proof It is an easy matter to check that $\{S_u : u \in I\!R^2\}$ is an $I\!R^d$-action on Ω equipped with Lebesgue measure (for the definition of $I\!R^d$-actions, see Chapter 2). In particular, each transformation S_u is measure preserving. Also, for $x \in I\!R^2$ and writing T_x for the translation in $I\!R^2$ over the vector x we have

$$P(X_A(B_1) = k_1, \ldots, X_A(B_n) = k_n)$$
$$= P(S_x^{-1}(X_A(B_1) = k_1, \ldots, X_A(B_n) = k_n))$$
$$= P(X_A(T_{-x}B_1) = k_1, \ldots, X_A(T_{-x}B_n) = k_n)$$

and hence the stationarity of X_A follows from the measure-preservingness of the group $\{S_u : u \in I\!R^2\}$.

For ergodicity we need to show that $\{S_u : u \in I\!R^2\}$ acts ergodically on Ω. This can be done as follows. Suppose that there exists a set $E \subset \Omega$ with $0 < P(E) < 1$ and which is invariant under all transformations $\{S_u : u \in I\!R^2\}$. This would imply that $P(S_u^{-1}E) = P(E)$ for all u. According to Lebesgue's

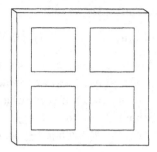

Figure 7.2. The generalised cutting and stacking.

density theorem, for all $\epsilon > 0$ it is the case that for almost all $x \in E$, there exists a k large enough so that the box $B_{x,k}$ of the form $[l_1, l_1 + 4^{-k}] \times [l_2, l_2 + 4^{-k}] \times [l_3, l_3 + 4^{-k}]$ containing x and where l_i is of the form $n_i 4^{-k}$ for integers n_i satisfies $\ell(E \cap B_{x,k})/\ell(B_{x,k}) \geq 1 - \epsilon$. This implies that for k large, the set Ω_k contains two little cubes with side length 4^{-k} such that E covers the first cube more than half, and E^c covers the second cube more than half. Now note that the 'thickness' of Ω_k is exactly 4^{-k}. Thus we can find a transformation S_u such that $S_u(E^c) \cap E$ has positive measure, a contradiction. $\qquad\square$

Here are some examples of stationary and ergodic point processes obtained this way.

Example 7.1   If the set A is a countable collection of points, then X_A satisfies $X_A(\mathbb{R}^d) = 0$ a.s. This is certainly a stationary and ergodic point process.

Example 7.2   Suppose that A is a set of the form $A = (x_1, x_2) \times [0, 1]$ for some $0 \leq x_1, x_2$. The point process obtained this way is just the point process of Example 1.1 in Section 1.3. As already explained in Chapter 1, we can think of this process as the integer lattice $\mathbf{Z}^2$ translated over a random vector which is uniformly distributed over the unit square.

The construction up to this point is not flexible enough for our purposes. Here is a generalisation: Choose a sequence $\{\alpha_n\}$ of positive numbers with the requirement that α_n grows at most polynomially in n. (This assumption is too strong but enough for our purposes.) Instead of rearranging the slices next to each other as above, we now construct a 'frame' around the slices of width α_n at the n-th iteration; see Figure 7.2. To this end, we define for each n mutually disjoint subsets $\Gamma_n \subset \mathbb{R}^3$ which are also disjoint from Ω. We shall see in a moment that we can take each Γ_n to be a finite union of disjoint blocks of a certain size. The stacking function f_1 in the first construction is replaced by a map $g_1 : \Omega \cup \Gamma_1 \to I\!\!R^3$, which on Ω is defined as

$$g_1(A_1^0) = [0, 1) \times [1, 2) \times [0, \tfrac{1}{4}) + (\alpha_1, 1 + 2\alpha_1, 0)$$

$$g_1(A_1^1) = [1, 2) \times [1, 2) \times [0, \tfrac{1}{4}) + (1 + 2\alpha_1, 1 + 2\alpha_1, 0)$$

$$g_1(A_1^2) = [0, 1) \times [0, 1) \times [0, \tfrac{1}{4}) + (\alpha_1, \alpha_1, 0)$$

$$g_1(A_1^3) = [1, 2) \times [0, 1) \times [0, \tfrac{1}{4}) + (1 + 2\alpha_1, \alpha_1, 0).$$

At this point we choose Γ_1 in such a way that we can define g_1 on Γ_1 such that g_1 is measure preserving and

$$g_1(\Gamma_1) = \{[0, 2 + 3\alpha_1]^2 \times [0, \tfrac{1}{4})\} \backslash g_1(\Omega).$$

One can think of g_1 acting on Γ_1 as taking blocks from Γ_1 and rearranging them so as to form a frame of width α_1 around the four slices coming from Ω. As such g_1 is still piecewise linear on $\Omega \cup \Gamma_1$. On $\cup_{i=2}^{\infty}\Gamma_i$, which will be defined in a moment, we shall define g_1 to be the identity. We write $\tilde{\Omega}_1 := g_1(\Omega \cup \Gamma_1) = [0, 2 + 3\alpha_1]^2 \times [0, \tfrac{1}{4})$. As before, if both (u, x) and $(u + u', x)$ are in $\tilde{\Omega}_1$ we set

$$S_{u'}(g_1^{-1}(u, x)) = g_1^{-1}(u + u', x) \tag{7.14}$$

as in (7.10). This procedure is repeated. We cut $\tilde{\Omega}_1$ into four slices which we rearrange with a frame of width α_2 around it, where the frame is constructed from a rearrangement of blocks which form the set $\Gamma_2 \subset I\!\!R^3$. The map which accomplishes this stacking is the analogue of stacking function f_2 above and is denoted by g_2. Hence g_2 is a piecewise linear map from $\tilde{\Omega}_1 \cup \Gamma_2 \to \tilde{\Omega}_2$. On $\cup_{i=3}^{\infty}\Gamma_i$, we define g_2 to be the identity. If (u, x) and $(u + u', x)$ are both in $\tilde{\Omega}_2$ we set

$$S_{u'}((g_2 \circ g_1)^{-1}(u, x)) = (g_2 \circ g_1)^{-1}(u + u', x) \tag{7.15}$$

as before. We continue in the obvious way, obtaining maps $g_n : \tilde{\Omega}_{n-1} \cup \cup_{i=n}^{\infty}\Gamma_i \to \tilde{\Omega}_n$ as analogues of the stacking maps f_n above. We put $\Gamma := \cup_{n=1}^{\infty}\Gamma_n$ and the probability space on which we define our point processes is just $\Omega \cup \Gamma$ with normalised Lebesgue measure P as to have $P(\Omega \cup \Gamma) = 1$. Note that it is here where we use the fact that the sequence $\{\alpha_n\}$ does not grow too fast: it is an easy matter to check that the Lebesgue measure of Γ is finite under our assumptions, i.e. the total volume of the framework added is finite. As before, for almost all $\omega \in \Omega \cup \Gamma$, the map $S_u(\omega)$ is defined for all $u \in I\!\!R^2$.

Ergodic point processes X_A, for $A \subset \Omega \cup \Gamma$ for which (7.12) is finite for every measurable and bounded set $B \subset I\!\!R^2$ can now be defined as in (7.13). Examples of this construction will be given in the next section.

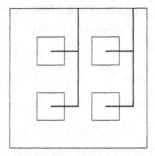

Figure 7.3. The bold line segments form the set $\pi(B_1)$. The large square is the front face of $\tilde{\Omega}_n$.

7.7 Examples

The cutting and stacking procedure of the previous section can be used as a 'counter-example machine'. We shall construct some examples which can be used to show that certain conditions in results of this chapter can not be omitted. We freely use the notation of the previous section.

Example 7.3 (A stationary tree)   Consider the generalised cutting and stacking procedure of the previous section and take $\alpha_n = 1$ for all n. If we define d_n via $\tilde{\Omega}_n = [0, d_n]^2 \times [0, (\frac{1}{4})^n)$, we see that d_n satisfies the recurrence relation $d_{n+1} = 2d_n + 3$, which is readily solved under the boundary condition $d_0 = 1$, giving that $d_n = 2^{n+2} - 3$ for all n. In order to describe a point process X_A we need to specify a set A. In fact, in order to specify A it is enough to specify sets of the form $g_n \circ \cdots \circ g_1(A) \cap \tilde{\Omega}_n =: A_n$ for all n and this is what we shall do. It is convenient to first describe a set $B_n \subset \tilde{\Omega}_n$ and then explain how A_n can be obtained from it. The first thing to remark is that if π denotes projection on the plane $\{(x_1, x_2, x_3) : x_3 = 0\}$, B_n and A_n will be such that $\pi^{-1}\pi(B_n) \cap \tilde{\Omega}_n = B_n$ and $\pi^{-1}\pi(A_n) \cap \tilde{\Omega}_n = A_n$ respectively. It is therefore enough to describe the projections $\pi(B_n)$ and $\pi(A_n)$ for all n. A picture is worth more than a thousand words, so the set $\pi(B_1)$ is defined to be the union of a finite number of straight line segments and is depicted in Figure 7.3.

The next iteration gives the set B_2 and $\pi(B_2)$ is depicted in Figure 7.4. It should be clear by now how the sets B_n for $n \geq 3$ are defined. This procedure yields a stationary (and ergodic) imbedding of a tree-like structure in the plane. The set $\pi(A_n)$ can now be obtained from $\pi(B_n)$ by replacing the straight line

Figure 7.4. The bold line segments form the set $\pi(B_2)$.

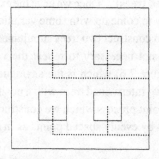

Figure 7.5. The set $\pi(A_1)$ is obtained from $\pi(B_1)$ by replacing the straight line segments by evenly spaced points.

segments which make up $\pi(B_n)$ by points evenly spaced having distance ϵ to each neighbouring point; see Figure 7.5 for the case $n = 1$.

The question may arise where the origin is in this construction and what the actual point process X_A looks like. Well, take n so large that $\omega \in \tilde{\Omega}_n$, and denote by 0 the projection $\pi(\omega)$ of ω. From the fact that $\pi^{-1}\pi(A_n) \cap \tilde{\Omega}_n = A_n$ we see that $S_x(\omega) \in A_n$ if and only if $S_x(0) \in \pi(A_n)$. So once $\omega \in \tilde{\Omega}_n$, the projections $\pi(A_n)$ tell it all and the points which make up $\pi(A_n)$ are the actual points of the point process X_A. Once ω is in $\tilde{\Omega}_n$, each further iteration determines the points of the point process in a larger box containing the origin.

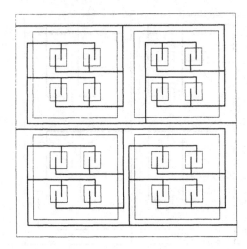

Figure 7.6. The second iteration in the construction of two disjoint stationary trees with a positive distance from each other.

Example 7.4 (Intertwined trees)   Once we have the example of the stationary tree above, it is quite easy to come up with some variations on the same theme. Instead of one tree we can construct two trees simultaneously in the same spirit as in Example 7.1. It is not necessary to repeat the whole construction, we just give the picture (Figure 7.6) which is the analogue of Figure 7.4. In this construction we obtain two intertwined trees with a positive distance from each other. Of course, in the point process which we construct from this, we replace all straight line segments by evenly spaced points as in the first example.

Example 7.5 (Multiple trees)   The construction in Example 7.4 cannot be extended to any finite number of trees. Here is a way to construct any finite number of stationary trees using the cutting and stacking procedure. We depict the first iteration of a construction in Figure 7.7, as it will be clear how to continue.

After these examples we discuss the relevance to percolation theory. From Theorem 7.4 we know that whenever the radii of the balls are unbounded, there can be at most one unbounded occupied component. Furthermore, in two dimensions, if the radii can be arbitrary small, there can be at most one unbounded vacant component (Theorem 7.5). In an RCM, we know from Theorem 7.8 that infinite range models have at most one unbounded component a.s. As already noticed, it is not hard to find examples of a Boolean model or RCM

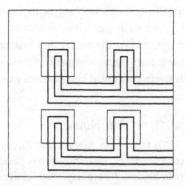

Figure 7.7. The first iteration in the construction of two stationary 'parallel' trees. It is clear how to generalise this to any finite number of trees.

where infinitely many unbounded components arise a.s. The point of the examples above is that they give rise to Boolean models or RCMs with multiple but finitely many unbounded components.

First, consider the construction in Example 7.5. Suppose that the distance between successive points on the trees is $\epsilon > 0$ and that the distance between the different trees is δ with $\delta > \epsilon$. Consider a Boolean model driven by this point process, and where the radius random variable ρ satisfies $P(\rho = \epsilon) = 1$. It is then clear that each tree in the construction gives rise to exactly one unbounded vacant component and also that the unbounded components corresponding to different trees are disjoint. Hence we have created a Boolean model with two unbounded occupied components. Generalisation to any finite number of unbounded components is clear. We remark that a Boolean model with fixed radii is in fact an RCM too. Thus this also gives an example of an RCM with two unbounded components.

A small variation of the construction in Example 7.4 may be used to create an example of a Boolean model where the radii are bounded from above but not from below, such that the model has two unbounded occupied components a.s. (and, according to Theorem 7.5, only one unbounded vacant component). The idea is the following. The trees have line segments of increasing lengths $l_1, l_2, \ldots$ say. Depending on the precise distribution of ρ we put, on each line segment with length l_n, so many points (evenly spaced) that the probability that the whole line segment is contained in the union of all balls centred at these points is at least $1 - 2^{-n}$. Starting at an arbitrary point on a tree there is a unique sequence of line segments along which we can radiate to infinity starting at that point. Denote this sequence by $l_{n_1}, l_{n_2}, \ldots$. From the Borel–Cantelli lemma it follows that with probability 1, all line segments l_{n_k} are covered by balls for k sufficiently large. Thus each tree gives rise to at least one unbounded component. A little thought reveals the conclusion that each tree can give rise to at most one unbounded component and we conclude that, in this Boolean model, exactly two unbounded occupied components exist a.s.

7.8 Notes

The material in Section 7.2 is taken from Meester and Roy (1994). The topological structure of unbounded components in two dimensions is based on Burton and Keane (1989, 1991). The results in Section 7.4 are improved versions of results in Meester and Roy (1994), and the results in Section 7.5 are taken from Burton and Meester (1993). Cutting and stacking goes back to Rudolph (1979), and the examples constructed using this technique are from Meester and Roy (1994).

8

Other continuum percolation models

This chapter is devoted to a miscellany of random processes related to the models discussed so far. Some of these models have been studied quite extensively but others are new. Consequently, there are many open questions in these models and we hope that this small survey will initiate research in this direction.

8.1 Continuum fractal percolation

There is a natural way to construct fractal like sets with countably many Boolean models. Take $\lambda > 0$ and consider the two-dimensional Boolean models $(X_k, 2^{1-k}, 4^{k-1}\lambda)$, for $k = 1, 2, \ldots$. The reason for this particular choice of the parameters will become clear as we proceed, but note at this stage that the covered volume fraction (CVF) of all these Boolean models is the same. (For the definition of the CVF, see Chapter 5.) Now denote by V_n the vacancy which remains after the superposition of the first n Boolean models. We denote this superposition by (Y_n, ρ_n, λ_n). Obviously, we have $V_1 \supseteq V_2 \supseteq V_3 \supseteq \cdots$ and we define the limit by

$$V_\infty = \bigcap_{k=1}^{\infty} V_k.$$

It is not at all clear at first sight that V_∞ can possibly be non-empty. Indeed, the CVF of a model is equal to the probability that a particular point in the plane is covered. If we denote the CVF of an individual Boolean model $(X_k, 2^{1-k}, 4^{k-1}\lambda)$ by α, then the probability that the origin is in V_n is equal to $(1 - \alpha)^n$; see Chapter 5. Thus the probability that the origin is in V_∞ is zero. But now we cannot, as in the proof of Proposition 3.1, conclude that the whole plane is covered a.s. The point is that this time infinitely many balls are centred in the unit square a.s. and as such we do not have local finiteness and

209

the argument in the proof of Proposition 3.1 breaks down. In fact, as we shall see now, if λ is sufficiently small, the vacant region is *not* empty a.s.

Proposition 8.1 *If $\lambda > 4 \log 4$, then $V_\infty = \emptyset$ a.s.*

Proposition 8.2 *If $\lambda < \dfrac{\log 4}{45}$, then $V_\infty \neq \emptyset$ a.s.*

Proof of Proposition 8.1 The proof proceeds by a simple branching process argument. Consider the unit square I^2, say, and divide I^2 into four smaller squares with side length $\frac{1}{2}$. The 0-th generation is just I^2, and the first generation of our branching process are those squares S among the four subsquares for which $X_1(S) = 0$. Note that whenever $X_1(S) \geq 1$, the whole square S is covered. Hence the union of the squares in the first generation contains the vacant region in $(X_1, \lambda, 1)$. Note also that the probability that a particular subsquare is in the first generation is equal to $\exp(-\lambda/4)$.

Each of the subsquares in the first generation is now divided into four further subsquares with side length $\frac{1}{4}$. Such a further subsquare S is in the second generation if and only if $X_2(S) = 0$. This happens with probability $\exp(-4\lambda/16) = \exp(-\lambda/4)$. It is clear that different squares in the first generation give birth to members in the second generation independently of each other. Thus we have constructed a branching process in such a way that whenever this process becomes extinct, there is no vacancy left in the original model. Extinction takes place a.s. whenever the expected number of members in the first generation is less than 1; i.e. if $4 \exp(-\lambda/4) < 1$, i.e. if $\lambda > 4 \log 4$. $\qquad\square$

Proof of Proposition 8.2 This result can also be proved by a branching process argument. This time we shall construct a branching process in such a way that if this process survives, then $V_\infty \neq \emptyset$. To this end, consider again the unit square I^2 and suppose that $X_1(I^2) = X_2(I^2) = 0$. This happens with positive probability. As before, the 0-th generation of our branching process consists of I^2 only. The first generation consists of those squares among $[0, \frac{1}{4}] \times [0, \frac{1}{4}]$, $[0, \frac{1}{4}] \times [\frac{3}{4}, 1]$, $[\frac{3}{4}, 1] \times [0, \frac{1}{4}]$ and $[\frac{3}{4}, 1] \times [\frac{3}{4}, 1]$ (i.e. all 'corner subsquares' of I^2) which are not intersected by any ball coming from X_3 or X_4. The probability that e.g. $[0, \frac{1}{4}] \times [0, \frac{1}{4}]$ is in the first generation is at least as large as the probability that there is no point of X_3 and X_4 in $[-\frac{1}{4}, \frac{1}{2}] \times [-\frac{1}{4}, \frac{1}{2}]$. This probability is equal to $\exp(-16\lambda\frac{9}{16}) \exp(-64\lambda\frac{9}{16}) = \exp(-45\lambda)$. The other corner squares are in the first generation with the same probability.

Now each of the squares of the first generation is divided into 16 subsquares and the second generation consists of those 'corner squares' among these which are not intersected by any ball coming from X_5 or X_6. The probability that this happens is the same as in the corresponding event described above. Also note that members of the first generation give birth to members of the second generation independently of each other. We continue in the obvious way. Now observe that if this branching process survives, we have a non-increasing sequence $A_1 \supseteq A_2 \supseteq \cdots$ of non-empty compact sets such that $A_n \subseteq V_n$ for all n. This implies that $V_\infty \supseteq \cap_{n \geq 1} A_n \neq \emptyset$ by Cantor's intersection theorem. Survival is possible with positive probability if $4 \exp(-45\lambda) > 1$, i.e. $\lambda < \log 4/45$. A reader concerned with details will notice that in the statement of the proposition we claim that V_∞ is not empty a.s. To see that this stronger statement is also true, we divide the plane into unit squares and consider a sequence $I_1, I_2, \ldots$ of such squares such that $d(I_j, I_k) > 4$ for all $j \neq k$ (to guarantee independence). Now $P(V_\infty \cap I_j \neq \emptyset)$ is positive and independent of j whence $P(V_\infty \neq \emptyset) = 1$. □

From a percolation point of view we are interested in the existence of large connected components of V_∞. So our next task is to show that large components do indeed exist. To this end, we define a new critical density λ_f (the 'f' refers to 'fractal') as follows:

Definition 8.1 Let $\theta_f(\lambda)$ be the probability that $V_\infty \cap [0, 1]^2$ contains a connected component which intersects the left and right sides of $[0, 1]^2$. Define λ_f as

$$\lambda_f = \inf\{\lambda \, : \, \theta_f(\lambda) = 0\}.$$

Note that λ_f is defined in terms of crossing probabilities like λ_S and λ_S^* in Chapters 3 and 4. We shall see later that λ_f is strongly related to the classical critical densities in the ordinary Poisson Boolean model. Our first task is to show that λ_f is not equal to zero.

Theorem 8.1 $\lambda_f > 0$.

Proof The idea of the proof is related to the proof of the well-known extinction theorem for branching processes. First we define the notion for the unit square I^2 to be m-*good*, for all $m \geq 0$. We assume that $X_1(I^2) = X_2(I^2) = X_3(I^2) = X_4(I^2) = 0$, something which happens with positive probability. We divide the unit square into 256 subsquares which we shall call *level-1 squares*. We say that I^2 is 0-good if at least 255 of these level-1 squares do not contain any point of X_5, X_6, X_7 and X_8. A level-1 square which does not contain any such point will be called *empty*. We further divide each empty level-1 square into 256 level-2 squares. An empty level-1 square is called 0-good if at least 255 of its level-2 subsquares do not contain any point of X_9, X_{10}, X_{11} and X_{12}. The unit square is said to be 1-good if it contains at least 255 0-good level-1 squares. Inductively, the unit square is called m-good if at least 255 of its 256 level-1 subsquares are $(m-1)$-good. We denote the probability that the unit square is m-good by $\theta_m(\lambda)$.

The idea behind these definitions is the following. If the unit square is 0-good, then it is an easy matter to check that each side of the unit square touches at least five level-1 squares which are completely vacant (with respect to Y_8). (Note that the diameter of the balls associated with X_5 is $\frac{1}{16}$.) The same statement, properly scaled, is true for 0-good level-1 squares. Two squares are said to be *adjacent* if they share a side. A *path* of squares is a finite sequence of squares $S_1, S_2, \ldots, S_k$ such that S_i and S_{i+1} are adjacent for all $i = 1, \ldots, k-1$. If both $[0, 1]^2$ and $[1, 2] \times [0, 1]$ are 0-good, then there exists a connection from $\{0\} \times [0, 1]$ to $\{2\} \times [0, 1]$ (inside the rectangle) of adjacent vacant (w.r.t. Y_8) level-1 squares. It is now easy to show inductively that if the unit square is m-good, then there is a path of vacant (with respect to $Y_{4(m+2)}$) level-m squares connecting the left and right sides of the unit square. Thus if I^2 is m-good for all $m = 1, 2, \ldots$, then we can find a non-increasing sequence of connected compact sets A_m with the properties that (i) $A_m \subseteq V_m$, and (ii) A_m intersects the left and right sides of I^2. It follows immediately that $V_\infty \neq \emptyset$ in such a case. It therefore suffices to show that if λ is sufficiently small then

$$\lim_{m \to \infty} \theta_m(\lambda) > 0. \tag{8.1}$$

To prove this, let $p := \exp(-85\lambda)$ be the probability that a level-m square is empty (i.e. does not contain a point of X_{4m+1}, X_{4m+2}, X_{4m+3} and X_{3m+4}.). Writing θ_m as a function of p rather than of λ we obtain by definition:

$$\theta_m(p) = 256p^{255}(1-p)(\theta_{m-1}(p))^{255} + \\ + p^{256}(256(\theta_{m-1}(p))^{255}(1-\theta_{m-1}(p)) + (\theta_{m-1}(p))^{256},$$

for $m \geq 1$, and

$$\theta_0(p) = p^{256} + 256p^{255}(1-p).$$

Define $\psi_p(x) = p^{255}x^{255}(256 - 255px)$. It then follows that

$$\theta_m(p) = \psi_p^{m+1}(1),$$

for all $m \geq 0$. It is easy to check that $\psi_p(x)$ is increasing in both p and x. It follows that $\lim_{m \to \infty} \theta_m(p)$ is equal to the largest fixed point of ψ_p in $[0, 1]$. We need only to show now that for p sufficiently large but smaller than 1 (which means for λ sufficiently small but positive) the largest fixed point of ψ_p in $[0, 1]$

is larger than zero. This, however, follows from the fact that $\psi_1(1) = 1$ and $(d/dx)\psi_1(x)|_{x=1} = 0$. □

We shall now formulate a relation between 'ordinary' Poisson Boolean models and the fractal model of this section. Fix some $n \geq 0$ and consider the (independent) Poisson Boolean models

$$(X_{1+(i-1)n}, 2^{-(i-1)n}, 4^{(i-1)n}\lambda),$$

for $i = 1, 2, \ldots$, defined on the same probability space. The i-th model will be denoted by $Z_i(n)$ and when there is any chance of confusion the probability measure in that model will be denoted by $P_{Z_i(n)}$. Note that $Z_1(n)$ is just $(X_1, 1, \lambda)$ for all n. Also observe that $Z_{i+1}(n)$ can be obtained from $Z_i(n)$ by scaling with a factor 2^{-n}. We shall denote the vacant region in $Z_i(n)$ by $V_i(n)$. Instead of looking at V_∞, we now concentrate on

$$\bigcap_{i=1}^{\infty} V_i(n) =: V(n).$$

The reason for doing so is that when n gets larger, $Z_1(n)$ will be more and more dominant so that $V(n)$ will be more and more like $V_1(1)$. But $V_1(1)$ is equivalent in law to the vacant region in the Boolean model $(X, 1, \lambda)$ and this will give the relation between the critical densities for fractal percolation and $\lambda_c^*(1)$. To make this precise, let $\theta_f^{(n)}(\lambda)$ be the probability of a vacant L–R crossing of the unit square in $V(n)$, and define

$$\lambda_f(n) = \inf\{\lambda : \theta_f^{(n)}(\lambda) = 0\}.$$

Theorem 8.2 $\lim_{n\to\infty} \lambda_f(n) = \lambda_c^*(1) \ (= \lambda_c(1))$.

Proof First we prove that

$$\lambda_f(n) \leq \lambda_c(1) \tag{8.2}$$

for all n. Recall that $\sigma((m, m), \lambda, 1)$ is the probability of an occupied L–R crossing of the square $[0, m]^2$ in the Poisson Boolean model $(X, 1, \lambda)$. Fix $\lambda > \lambda_c(1)$. From Corollary 4.1 we have that

$$\lim_{m\to\infty} \sigma((m, m), \lambda, 1) = 1. \tag{8.3}$$

Consider the event $E_i = \{$there is an L–R occupied crossing of the unit square in $Z_i(n)\}$. A simple scaling argument gives

$$P(E_i) = \sigma((2^{(i-1)n}, 2^{(i-1)n}), \lambda, 1).$$

From (8.3) and the independence of the events E_i we have that

$$P(\limsup_{i \to \infty} E_i) = 1.$$

By symmetry, this is also true if we consider T–B occupied crossings instead of L–R crossings. So infinitely many models $Z_i(n)$ (for fixed n) have occupied T–B crossings of the unit square a.s. and consequently $\theta_f^{(n)}(\lambda) = 0$.

The other inequality is harder, and we concentrate first on the ordinary Poisson Boolean model $(X, 1, \lambda)$. Fix some $\lambda < \lambda_c(1)$, let R_M be the rectangle $[0, 3M] \times [0, M]$ and let D_M be the square $[0, M]^2$. Using the FKG inequality and Corollary 4.1 we see that for all $\epsilon_1 > 0$, the following event $E_{M,\eta} = \{$there is a path of vacant squares of side length η crossing R_M from left to right and two such T–B crossings in D_M and $D_M + (2M, 0)$, respectively$\}$ satisfies

$$P(E_{M,\eta}) > 1 - \epsilon_1, \tag{8.4}$$

for M sufficiently large and η sufficiently small. Next we choose n so large that $2^{-n} < \eta$ and such that $2^n = (2k_n + 1)M$ for some integer k_n. We divide D_{2^n} into squares of side length M and we denote by $\mathbb{L}_{M,n}$ the set of vertices $\{(M/2, M/2) + (2Mi, 2Mj)\}$, where $i, j \in \mathbb{Z}$. We connect any two vertices v and v' in $\mathbb{L}_{M,n}$ if and only if $d(v, v') = 2M$. We now perform bond percolation on the ensuing lattice (also to be denoted by $\mathbb{L}_{M,n}$), declaring a bond to be open if the event $E_{M,\eta}$, properly translated and rotated, occurs in the union of the three squares which intersect the bond. If this is not the case, the bond is said to be closed. This is not an independent percolation model, but two bonds which do not have an end point in common are independent. Now let $F_{M,n}$ be the event that there is an L–R crossing of D_{2^n} in $\mathbb{L}_{M,n}$ of open bonds, i.e. a crossing from a vertex in $\{(i, j) \in \mathbb{L}_{M,n} : i = M/2\}$ to a vertex in $\{(i, j) \in \mathbb{L}_{M,n} : i = (2^n/M) - (M/2)\}$. It is clear from the construction that the occurrence of the event $F_{M,n}$ implies that there is an L–R vacant crossing of D_{2^n} in the underlying Poisson Boolean model $(X, 1, \lambda)$. To estimate $P(F_{M,n})$ we introduce the *dual* lattice $\mathbb{L}_{M,n}^d$ which is just the lattice $\mathbb{L}_{M,n}$ translated over the vector (M, M). Each bond in the dual lattice intersects one bond of the original lattice and a bond in the dual is declared open if and only if the intersecting bond is open, and closed otherwise. Now $F_{M,n}$ does not occur if and only if there a closed T–B crossing of $[M, 2^n - M] \times [-M, 2^n + M]$ in the dual (see Section 1.2 in Chapter 1). The probability of the latter event can be estimated by counting arguments as in Chapter 1: each such T–B crossing starts from any of the $(2^n - M)/(2M)$ vertices at the bottom and must contain at least $(2^n + M)/(2M)$ bonds. Furthermore, the number of distinct paths of length k is at most 3^k. Finally, the state of a bond in the dual depends on the state of only six other bonds in the dual and the geometry of this dependence structure is such that any path of length k in the dual contains at least $\lfloor k/4 \rfloor$ bonds which are mutually independent. Putting these observations together gives

$$P(F_{M,n}) \geq 1 - \frac{2^n - M}{2M} \sum_{k \geq \frac{2^n + M}{2M}} 3^k \epsilon_1^{k/4}$$

$$= 1 - \frac{2^n - M}{2M} \frac{(3\epsilon_1^{1/4})^{(2^n+M)/(2M)}}{(1 - 3\epsilon_1^{1/4})}. \tag{8.5}$$

Now we return to the fractal model. Recall that we have chosen $\lambda < \lambda_c(1)$. It suffices to show that for all n sufficiently large $V(n)$ contains an L–R crossing of the unit square with positive probability.

Assume that $[0, 1]^2$ is completely vacant in V_1. This happens with positive probability and is just for convenience. Scaling (8.4) and (8.5) yields

$$P_{Z_i(n)}(E_{M/2^{(i-1)n}, \eta/2^{(i-1)n}}) > 1 - \epsilon_1 \tag{8.6}$$

and

$$P_{Z_2(n)}(\text{there is a vacant L–R crossing of the unit square})$$

$$\geq 1 - \frac{2^n - M}{2M} \frac{(3\epsilon_1^{1/4})^{(2^n+M)/(2M)}}{(1 - 3\epsilon_1^{1/4})}$$

$$= 1 - h_{M,n}(\epsilon_1), \text{ say.} \tag{8.7}$$

The lattice construction above can also be carried out, suitably scaled, in any of the models $Z_i(n)$. Suppose vacant paths as in the definition of the event $E_{M/2^n, \eta/2^n}$ exist, and let $G_3(n)$ be the event that inside these paths, there is a path of vacant squares of side length $\eta/2^{2n}$ crossing $R_{M/2^n}$ from left to right after the model $Z_3(n)$ has been 'placed'. To estimate $P(G_3(n) \mid E_{M/2^n})$ we perform a similar discretisation as above, suitably scaled. For the counting argument, note that the path in the event $E_{M/2^n, \eta/2^n}$ consists of at most $3M^2$ squares. There are, on either side of the path, at most $3M^2[(2^n - M)/(2M)]$ vertices in the dual which are adjacent to an edge in the dual crossing an edge of the path. In order for $G_3(n)$ not to occur, one of these vertices in the dual has to be the starting point of a path of at least $(2^n + M)/(2M)$ closed edges. A similar calculation as above now yields

$$P(G_3(n) \mid E_{M/2^n, \eta/2^n}) \geq 1 - 3M^2 \frac{2^n - M}{2M} \frac{(3\epsilon_1^{1/4})^{(2^n+M)/(2M)}}{(1 - 3\epsilon_1^{1/4})}$$

$$= 1 - g_{M,n}(\epsilon_1), \text{ say.} \tag{8.8}$$

Hence from (8.6) we find

$$P(G_3(n)) \geq (1 - \epsilon_1)(1 - g_{M,n}(\epsilon_1))$$
$$\geq 1 - \epsilon_1 - g_{M,n}(\epsilon_1)$$
$$=: 1 - \epsilon_2.$$

So the probability that in the superposition of $Z_2(n)$ and $Z_3(n)$ there is a vacant crossing of $R_{M/2^n}$ by squares of side length 2^{-2n} is at least $1 - \epsilon_2$. This statement can be scaled properly so as to yield a similar statement about the superposition of $Z_i(n)$ and $Z_{i+1}(n)$ and suitable crossings in suitable rectangles. Define $G_k(n)$ as the event that in the superposition of $Z_2(n), Z_3(n), \ldots, Z_k(n)$ there exists a path of vacant squares of side length $2^{(k-1)n}$ crossing $R_{M/2^n}$ from left to right. Choosing

$$\epsilon_{k+1} := \epsilon_1 + g_{M,n}(\epsilon_k)$$

for all $k \geq 1$ we conclude that

$$P(G_k(n)) \geq 1 - \epsilon_{k-1} \tag{8.9}$$

for all $k \geq 2$. Thus the probability of a vacant L–R crossing of the unit square in $V_k(n)$, given that the unit square is contained in V_1 is bounded from below by the expression obtained from the right-hand side of (8.7) if we replace ϵ_1 by ϵ_k.

To complete the proof we again perform an iterative procedure. Take ϵ_1 such that $3(2\epsilon_1)^{1/4} < 1$. Now choose some M and η and choose n so large that $g_{M,n}(2\epsilon_1) < \epsilon_1$ and such that $h_{M,n}(2\epsilon_1) < 1$. (The reader may check easily that this can be done.) The function $g_{M,n}$ is non-decreasing and so is the function $\psi(x) := \epsilon_1 + g_{M,n}(x)$. Note that

$$\epsilon_{k+1} = \psi(\epsilon_k)$$

for all $k \geq 1$. From the choice of ϵ_1 we have that $\psi(\epsilon_1) > \epsilon_1$ and $\psi(2\epsilon_1) < 2\epsilon_1$, whence ψ has a fixed point in the interval $(\epsilon_1, 2\epsilon_1)$. It follows that $\epsilon := \lim_{k \to \infty} \epsilon_k$ exists and is contained in $(\epsilon_1, 2\epsilon_1)$. Hence $V(n)$ contains a vacant L–R crossing of the unit square with probability at least $1 - h_{M,n}(2\epsilon_1) > 0$. □

8.2 Percolation of level sets in random fields

Imagine a hilly landscape and a certain level h, say. The level h is supposed to represent the level to which the landscape has been filled with water. Typically, one expects that if h is sufficiently small, then there are only bounded lakes of water and an infinite land mass; if h is large enough, then there should be only bounded islands in an infinite ocean.

To formulate this model mathematically, we consider a stationary, ergodic, a.s. continuous random field $\{\psi(x) : x \in \mathbb{R}^d\}$. We shall always assume that $E\psi(x) = 0$ for all $x \in \mathbb{R}^d$. We define *level sets* as follows:

$$S_h = \{x : \psi(x) = h\},$$

and

$$S_{\leq h} = \{x : \psi(x) \leq h\},$$

for all $h \in \mathbb{R}$. As usual, we say that a subset of $\mathbb{R}^d$ *percolates* if it contains an unbounded connected component. In analogy with the percolation models discussed in this book we may define

$$h_c = h_c(\psi) = \inf\{h : S_{\leq h} \text{ percolates with positive probability}\}.$$

Note that if $S_{\leq h}$ percolates with positive probability it percolates almost surely by ergodicity.

First we shall give a condition under which $h_c(\psi)$ is bounded away from infinity. This is the analogue of Theorem 1.1 in Chapter 1 and, as we shall see, the proof proceeds very much along the same line. In order to formulate the condition for non-triviality of h_c, we discretise $\mathbb{R}^d$ in the usual way: the space is partitioned into unit cubes $B_1, B_2, \ldots$, where $B_i = z_i + (-\frac{1}{2}, \frac{1}{2}]^d$, for an enumeration $\{z_i\}$ of the vertices in $\mathbf{Z}^d$.

Theorem 8.3 *Suppose that there exists a non-increasing function $g : \mathbb{R} \to \mathbb{R}$ such that $g(h) \downarrow 0$ as $h \to \infty$ and a constant $c = c(h) > 0$ such that for any subset $\{B_{i_1}, B_{i_2}, \ldots, B_{i_k}\}$ of unit cubes*

$$P\left(\bigcap_{j=1}^{k}\left\{\max_{x \in B_{i_j}} \psi(x) \geq h\right\}\right) \leq c(h)\{g(h)\}^k. \tag{8.10}$$

Then $h_c(\psi) < \infty$.

Proof Consider the random field $\{\phi(z) : z \in \mathbf{Z}^d\}$ defined as

$$\phi(z_i) = \max_{x \in B_i} \psi(x).$$

We say that a vertex z is *h-open* if $\phi(z) \leq h$ and *h-closed* otherwise. If we can show that for h large enough, there is (discrete) site percolation of h-open vertices, then it follows from the a.s. continuity of ψ that $h_c(\psi) < \infty$.

It follows from (8.10) that for every finite subset $\{z_1, \ldots, z_k\}$ of vertices

$$P\left(\bigcap_{i=1}^{k}\{\phi(z_i) \geq h\}\right) \leq c(h)\{g(h)\}^k. \qquad (8.11)$$

To show that this implies that h-open percolation occurs in ϕ for h sufficiently large, we need to introduce the notion of so-called *-connections* in Z^d. This is the analogue of the dual graph in bond percolation (see Chapter 1). Two vertices z and z' are *-neighbours if $|z - z'| \leq \sqrt{d}$. Note that each vertex has $3^d - 1$ *-neighbours. We can define *-paths and *-clusters in the obvious way with this new connection rule. Now it can be seen that there is no h-open percolation if and only if there are infinitely many disjoint, h-closed *-connected sets 'surrounding' the origin. (Here 'surrounding' means that the origin is cut off from infinity from the percolation point of view.) Now we perform a counting argument as in the proof of Theorem 1.1. Let E_n be the event that there is a h-closed *-connected set of n vertices surrounding the origin. For each such set of n vertices, the probability that it is h-closed is at most $c(h)\{g(h)\}^n$, using (8.11). There are at most $n(3^d - 2)n - 1$ such sets whence $P(E_n) \leq n(3^d - 2)^{n-1}c(h)\{g(h)\}^n$. Now choose h so large that $g(h) < (3^d - 2)^{-1}$. Then $\sum_n P(E_n) < \infty$ and h-open percolation occurs a.s. □

This result may be applied to a number of special cases. The most obvious special case is a random field with *finite correlation radius*. This means that there exists an $R > 0$ such that for any collection $A_1, \ldots, A_n$ of bounded measurable sets with $\inf\{d(x, y) : x \in A_i, y \in A_j\} > R$ whenever $i \neq j$, the σ-algebras generated by the values of $\psi(x)$ on these sets are independent.

Corollary 8.1 *A stationary random field ψ with finite correlation radius has $h_c(\psi) < \infty$.*

Proof Let F be the distribution function of $\max_{x \in B_1} \psi(x)$. From the continuity of ψ we have that $F(x) \to 1$ as $x \to \infty$. Any collection of k disjoint unit cubes contains at least $k/(2R + 3)^d$ cubes whose σ-fields are independent. Now apply Theorem 8.3 with $c(h) = 1$ and $g(h) = (1 - F(h))^{1/(2R+3)^d}$. □

Another application of Theorem 8.3 can be found in the theory of stationary Gaussian random fields. If the correlations in such a field decay sufficiently fast, then it is possible to show that (8.10) holds. The correlation function R in a random field ψ is defined as $R(x) := E(\psi(0)\psi(x))$. We give the next result without proof.

Theorem 8.4 *Let ψ be a stationary Gaussian field and suppose there exists a non-increasing function $f : \mathbb{R} \to \mathbb{R}$ satisfying $\int_0^\infty x^{d-1} f(x)dx < \infty$ such that for some positive constants c_i*

(i) $|R(x)| \le f(|x|), |grad R(x)| \le c_1 f(|x|),$

(ii) $\left\| \dfrac{\partial^2 R}{\partial x_i \partial x_j} \right\| \le c_2 f(|x|),$

(iii) $|1 - R(x)| \le \log^{-(3+\delta)} \dfrac{1}{|x|}$*, for some $\delta > 0$ and for all $|x| < 1$.*

Then (8.10) is satisfied for suitable g and hence $h_c(\psi) < \infty$.

We continue the discussion with a different class of random fields in two dimensions. Let G denote either the square lattice or the triangular lattice (see the last section of Chapter 3) with bonds of length 1. Let $\{x_i : i = 1, 2, \ldots\}$ be the set of vertices of G. Let $\{A_i : i = 1, 2, \ldots\}$ be a sequence of i.i.d. random variables with zero mean. Finally, let $\phi : \mathbb{R} \to \mathbb{R}$ satisfy $\int_{\mathbb{R}} x\phi(x)dx < \infty$. The latter condition is just to guarantee that the model is non-trivial. Now define a random field ψ on $\mathbb{R}^2$ as

$$\psi(x) = \sum_{i=1}^\infty A_i \phi(|x - x_i - U|), \qquad (8.12)$$

where U is a random vector uniformly distributed over a particular face of G. (The vector U is only there to make sure that ψ is stationary and it has no effect on the important features of the realisations.) This type of random fields has received some attention in the physics literature; see the references in the Notes. Apart from questions concerning the non-triviality of $h_c(\psi)$, the behaviour in the subcritical regime has been an object of research. Here some interesting phenomena can occur. In Boolean models with bounded radii we showed that the phase transition is sharp in the sense that if there is no percolation, then the distribution function of the size of the components of the origin goes down exponentially fast. It turns out that this need not be the case here. To describe this, we specialise to the case in which $P(A_1 = 1) = P(A_1 = -1) = \frac{1}{2}$ and ϕ is a smooth, strictly decreasing and strictly convex function with support $[0, \frac{1}{2} + \epsilon]$ with $\phi(0) = 1$, where $\epsilon > 0$ is chosen such that balls centred at the sites of G with radius $\frac{1}{2} + \epsilon$ intersect only pairwise.

Theorem 8.5 *If G is the triangular lattice, we have*

$$h_c(\psi) = 2\phi(\tfrac{1}{2}),$$

and for all h with $|h| < h_c(\psi)$ *we have*

$$P(d(S_h) > t \mid \psi(0) = h) \geq c_1(h)t^{-\alpha},$$

for positive constants $c_1(h)$ *and* α *where* α *does not depend on h and* $d(\cdot)$
denotes diameter of a set. If G is the square lattice, we have

$$h_c(\psi) = 0,$$

and for all $h \neq 0$,

$$P(d(S_h) > t \mid \psi(0) = h) \leq c_2 e^{-c_3 t},$$

for positive constants c_2 *and* c_3.

The different behaviour for the triangular and square lattice is due to the different geometry of the lattices: in discrete percolation on the triangular lattice, a finite open cluster is surrounded by a closed circuit. On the square lattice, however, a finite open cluster is surrounded by a closed $*$-cluster, as noted in the proof of Theorem 8.3.

Partial proof of Theorem 8.5   Consider first the case where G is the triangular lattice. The critical probability for independent site percolation on the triangular lattice is $\frac{1}{2}$, and there is no percolation at criticality (see Kesten 1982) whence the origin is surrounded by infinitely many disjoint open and infinitely many disjoint closed circuits. We can perform independent site percolation by declaring the site x_i to be open if and only if $A_i = 1$ and closed otherwise. Note that it follows from the convexity of ϕ that for any bond b connecting two plus sites (i.e. sites x_i with $A_i = 1$), $\inf_{x \in b} \psi(x) = 2\phi(\frac{1}{2})$. This implies immediately that for $h < 2\phi(\frac{1}{2})$, $S_{\leq h}$ cannot cross an open circuit and hence does not percolate. Conversely, each face of the lattice contains a region where $\psi(x) = 0$, and it is not hard to see that if $h > 2\phi(\frac{1}{2})$, then any two such regions in adjacent faces are connected in $S_{\leq h}$. This implies that $h_c(\psi) = 2\phi(\frac{1}{2})$.

To prove the corresponding result when G is the square lattice, we note that it is known that the critical probability p_c for independent site percolation on the square lattice is strictly larger than $\frac{1}{2}$, and that the critical probability p_c^* for $*$-percolation satisfies $p_c + p_c^* = 1$ (see Notes). This means that there are infinitely many $*$-circuits of either type surrounding the origin, but only finitely many ordinary circuits of either type. Now let $h < 0$. Given a plus $*$-circuit we can, by transforming the bonds slightly so as to avoid balls centred at minus vertices, find a curve through the same faces and vertices as the $*$-circuit such that $\psi(x) \geq 0$ on the curve. The sets $S_{\leq h}$ cannot cross such curves and hence $S_{\leq h}$ does not percolate. Conversely, it is not hard to see that a component of S_0

can only be bounded if it is surrounded by a plus or minus circuit in the lattice. As observed above, there are only finitely many such circuits surrounding the origin and we conclude that S_0, and thus also $S_{\leq 0}$ percolates.

To give the idea behind the proof of the two remaining statements, note that $\frac{1}{2}$ is critical for independent site percolation on the triangular lattice, but subcritical for independent site percolation on the square lattice. As a result of this, if C denotes the cluster of the origin in either model (so depending on the state of the origin, C is a plus or a minus cluster), the function $P(|C| \geq n)$ goes down (when $n \to \infty$) only polynomially in the triangular lattice, but exponentially in the square lattice. The idea of the proof is now to relate the size of the level set of the origin to the cluster C in the coupled discrete percolation model. For the square lattice this is quite simple, as for $h \neq 0$, the level set $S_{\psi(0)}$ which contains the origin is contained in $\cup_{x_i \in C_0} S(x_i, \frac{1}{2} + \epsilon)$. For the triangular lattice, the proof is a little more involved and we do not give it here. (See references in the Notes.) □

8.3 Dependent Boolean and random-connection models

In the standard Poisson Boolean model, each point of a Poisson point process X with density $\lambda > 0$ is the centre of a ball with random radius. Radii of different balls are independent of each other and all radii are independent of X. In this section, we introduce a stationary (and ergodic) model where the radii are no longer independent of each other and the point process.

We start with a Poisson process in $I\!R^d$ with density λ. In the model the density turns out to be irrelevant (as can be seen by a simple scaling argument) and we take it to be equal to 1. Choose an integer $k \geq 1$, the parameter of the model. The configuration of balls in space is constructed dynamically as follows. At time 0, all points of X are the centre of a ball with radius 0. Then, as time t evolves, the radius of each ball grows linearly in t, and all radii grow with the same speed. Balls start intersecting each other while growing and each ball remembers with how many balls it has non-empty intersection. As soon as a ball hits the k-th ball, it stops growing forever. Thus at each time t the space is partitioned into a region which is occupied by balls and its complement which we call the vacant region. Let $C_t^d(k)$ be the occupied region at time t. We are interested in the limiting configuration $C^d(k)$ defined as

$$C^d(k) = \bigcup_{t \geq 0} C_t^d(k).$$

Note that it is not completely obvious that this model exists, in the sense that $C_t^d(k)$ can actually be constructed this way for all values of t. The problem is that at any time, a ball might need 'information from infinity' to decide whether or not it can continue growing. Actually, the argument in Case 2 in the proof of Theorem 8.6 below can easily be modified as to obtain an existence proof. We do not elaborate this here and refer to the references in the Notes.

From a percolation point of view we are interested in the existence of unbounded connected components in $C^d(k)$. If these exist, we say that $C^d(k)$ percolates. Let us define the critical k as $k_c(d) := \min\{k \geq 1 : C^d(k)$ percolates with positive probability$\}$. The parameter space in this model is discrete and this gives us some hope that $k_c(d)$ might be computed explicitly. We shall prove the following estimate:

Theorem 8.6 *For all $d \geq 2$, it is the case that*

$$2 \leq k_c(d) < \infty.$$

The fact that $k_c(d) < \infty$ follows from Theorem 8.7 below. Therefore we only prove the first inequality here. From now on, $k = 1$ and we shall prove that $C^d(1)$ does not percolate a.s. The argument will be dimension free, so we write $C := C^d(1)$ and $C_t := C_t^d(1)$ from now on.

It will be convenient to define a graph T as follows. The vertices of T are the occurrences of X and two points x and y are neighbours (to be denoted by $x \sim y$) if the balls centred at x and y are tangent (or, equivalently, have non-empty intersection). To say that C percolates is the same as to say that T contains an infinite component (in the usual graph-theoretical sense). We say that the point x is smaller than y (notation: $x \prec y$) if the ball centred at x has smaller radius than the ball centred at y. We define the relation '$\preceq$' between points of X in the obvious way.

Lemma 8.1 *With probability 1, each point x of X has at most one neighbour y for which $y \preceq x$.*

Proof   The only way for a ball to get a neighbour with the *same* radius is to hit each other while both are still growing. This obviously implies that a ball can have at most one such neighbour and in such case has no smaller neighbours.

The only way for a ball to get a *smaller* neighbour is to hit a ball which already stopped growing before. Hence it suffices to show that it is a.s. impossible that a growing ball hits two or more balls simultaneously. This is quite obvious and one way of seeing this is the following. Select two points x and y of X and wait until they both stop growing. Suppose this happens at time t_0. Consider the union of the components in C_{t_0} containing x and y and denote this union by W. Given W, the point process X outside the region $W' := \{x \in \mathbb{R}^d : |x - W| \leq t_0\}$ is still unconditioned. But all potential balls which might hit W at two different balls at the same time are centred outside W'. Note that W consists of the union of finitely many balls and the set of points outside W' which have the same distance to two or more balls of W has Lebesgue measure zero. Hence the probability that the balls associated with x and y are hit simultaneously by a larger ball is zero. □

It follows from Lemma 8.1 that T is almost surely a forest, i.e. its components are a.s. trees. To see this, note that if T contains a circuit, then this circuit has to contain a largest point (i.e. a point with largest associated radius), which leads to a contradiction if we consider the two neighbours of this point in the circuit. Furthermore, two tangent balls in C have the same radius if and only if they stop growing at the same time, i.e. when they hit each other. We call two such balls a *root*. It follows from Lemma 8.1 that any component in C can contain at most one root a.s. To see this, suppose there is a component with two roots. It is obvious that the two roots have different associated radii a.s. The balls in the larger root can, according to Lemma 8.1, only have larger neighbours and such a neighbour can again only have larger farther neighbours and so on. Hence a path to the smaller root cannot exist a.s.

Proof of Theorem 8.6   We shall derive a contradiction by assuming that C percolates with positive probability. An unbounded component in C either contains a root or does not contain one. We rule out both possibilities separately:

CASE 1: Suppose that with positive probability (and hence with probability 1 by ergodicity) C contains an unbounded component W with a root. As remarked above, W contains exactly one root a.s. in such a situation. We call the point of intersection between two balls of a root contained in an unbounded component an *encounter point*. If encounter points exist, then there has to be a density $\mu > 0$, say, of such points in space. Then, for all $K > 0$ there exists a number N_K such that the following event E has probability at least $\mu/2$: $E := \{$the unit box $B_{1/2}$ contains an encounter point and the associated unbounded component contains at least K points in $B_{N_K}\}$, where B_n denotes the box $[-n, n]^d$. For $z \in \mathbf{Z}^d$, the event $E(z)$ is defined by replacing $B_{1/2}$ and B_{N_K} by $z + B_{1/2}$ and $z + B_{N_K}$, respectively. It follows from the ergodic theorem that for M sufficiently large (depending on the realisation) the box B_M contains at least $(\mu/4)(2M)^d$ cubes of the form $z + B_{1/2}$ for which $E(z)$ occurs and for which $z + B_{N_K} \subseteq B_M$. However, the sets of K points associated with the different encounter points are mutually disjoint whence

$$X(B_M) \geq \frac{\mu}{4}(2M)^d.$$

On the other hand, the ergodic theorem implies that for all M large enough

$$X(B_M) \leq 2(2M)^d$$

(remember that $\lambda = 1$ throughout). Taking $K > 8\mu^{-1}$ now gives the required contradiction.

CASE 2: Next we rule out the possibility of unrooted unbounded components. First note that an unrooted component cannot contain a smallest point as this point would be one of a root. Also, a point cannot have only neighbours which are all strictly larger than the point itself. Hence every point has at least one neighbour which is strictly smaller than the point itself, and we conclude that any unbounded unrooted component contains an infinite sequence of tangent balls with strictly decreasing radii. The radii in such a sequence approach a limit α, say, which is random. Note however that the set Γ of possible limits is non-random by ergodicity. We now show that Γ has to be empty.

First suppose that $0 \in \Gamma$. This would imply that for every $\epsilon > 0$, the standard Poisson Boolean model with balls of fixed radius ϵ percolates for $\lambda = 1$. However, for ϵ sufficiently small $\lambda = 1$ is subcritical (see Chapter 3) and we have a contradiction. Next suppose that for some $\beta > 0$ we have $\beta \in \Gamma$. We can assume that there is a Poisson point at the origin. Then, for all $\epsilon > 0$ there is a positive probability that the radius of the ball at the origin is in $(\beta, \beta + \epsilon)$ and that this ball is one of an infinite chain of tangent balls with decreasing radii which are all at least β. We shall now prove with a branching process argument that this is impossible. Denote the ball centred at x with radius r by $S(x, r)$. First, we choose $\epsilon > 0$ so small that the annulus $A(2\beta, 2\epsilon) := S(0, 2\beta + 2\epsilon) \backslash S(0, 2\beta)$ has d-dimensional Lebesgue measure less than 1. We are going to construct a Poisson process with density 1 in $I\!R^d$ step by step as follows. First consider a Poisson process X_1 with density 1 restricted to $A(2\beta, 2\epsilon)$. The expected number of points of X_1 is at most one by construction. If there are no points we stop, otherwise denote the points by $x_1, \ldots, x_n$. Now we concentrate on x_1 first and consider a Poisson process X_2 (independent of X_1) with density 1 in $(x_1 + A(2\beta, 2\epsilon)) \backslash A(2\beta, 2\epsilon)$. The points of X_2 are denoted by $x_{1,1}, \ldots, x_{1,n_1}$. As before, the expected number of such points is less than 1. Next, we examine the point x_2 and put a Poisson process X_3 with density 1 in $x_2 + A(2\beta, 2\epsilon) \backslash (A(2\beta, 2\epsilon) \cup (x_1 + A(2\beta, 2\epsilon)))$ and denote the points of this process by $x_{2,1}, \ldots, x_{2,n_2}$. We continue in the obvious way, each time adding a Poisson process in a region which is disjoint from all regions inspected so far. It is clear from the construction that if this branching process dies out, the origin cannot be in a chain with decreasing radii all of which are at least β. But by construction, the branching process dies out a.s. and the proof is complete. □

We continue with a 'dependent RCM' in the same spirit as the previous example. The setup is the same: take a Poisson process in $\mathbb{R}^d$ with density 1 (as in the first model of this section, the density is irrelevant). Now we connect each point x of X with the m points of X nearest to x. Again, $m \geq 1$ is the parameter of the model and we can define $m_c(d)$ as the smallest m for which percolation occurs in this model. For $m_c(d)$, we have the same bounds as for $k_c(d)$ above:

Theorem 8.7 *For every $d \geq 2$ it is the case that*

$$2 \leq m_c(d) < \infty.$$

The proof of the lower bound in Theorem 8.7 proceeds by a branching process argument as in the proof of Theorem 8.6 above and we do not give it here. We shall now show that $m_c(d)$ is bounded away from infinity. The bound we obtain is very crude and the conjecture, based on simulations, is that $m_c(2) = 3$ and $m_c(d) = 2$ for all $d \geq 3$.

Proof of Theorem 8.7 For ease of exposition, we will give the proof for the case $d = 2$ only, the generalisation to higher dimensions being completely straightforward. Let p_c denote the critical value for site percolation on the square lattice. As noted before, the density λ of the Poisson process is irrelevant for the occurrence of infinite clusters, so we can pick λ so large that the probability of seeing no point in the square $[0, \frac{1}{7}]^2$ satisfies

$$P\left(X\left(\left[0, \frac{1}{7}\right]^2\right) = 0\right) < \frac{1 - p_c}{2 \cdot 7^2}.$$

Let $E'_{0,0}$ be the event that for $i, j = 0, \ldots, 6$ we have that

$$X\left(\left[\frac{i}{7}, \frac{i+1}{7}\right] \times \left[\frac{j}{7}, \frac{j+1}{7}\right]\right) > 0;$$

i.e., $E'_{0,0}$ is the event that we see at least one point in each of the 7^2 basic subsquares (of the form $[i/7, (i+1)/7] \times [j/7, (j+1)/7]$) of the unit square

$[0, 1]^2$. We have

$$P(E'_{0,0}) > 1 - 7^2 \left(\frac{1 - p_c}{2 \cdot 7^2}\right) = \frac{1 + p_c}{2}.$$

Now pick m so large that the probability of seeing more than $m/7^2$ points in the square $[0, \frac{1}{7}]$ satisfies

$$P\left(X\left(\left[0, \frac{1}{7}\right]^2\right) > \frac{m}{7^2}\right) < \frac{1 - p_c}{2 \cdot 7^2}.$$

Let $E^*_{0,0}$ be the event that for $i, j = 0, \ldots, 6$ we have

$$X\left(\left[\frac{i}{7}, \frac{i+1}{7}\right] \times \left[\frac{j}{7}, \frac{j+1}{7}\right]\right) \leq \frac{m}{7^2},$$

i.e., that we see at most $m/7^2$ points of X in each of the 7^2 basic subsquares. We have

$$P(E^*_{0,0}) > 1 - 7^2 \left(\frac{1 - p_c}{2 \cdot 7^2}\right) = \frac{1 + p_c}{2}.$$

Let $E_{0,0}$ be the event given by $E_{0,0} = E'_{0,0} \cap E^*_{0,0}$, and for $l, n \in \mathbf{Z}$, let $E_{l,n}$ be the obvious analogous event for the square $[l, l+1] \times [n, n+1]$. We have that the events $\{E_{l,n}\}_{l,n \in \mathbf{Z}}$ are independent with probabilities

$$P(E_{l,n}) = P(E_{0,0}) > 1 - (1 - P(E'_{0,0})) - (1 - P(E^*_{0,0})) = p_c.$$

For $i = 0, \ldots, 7$, let S_i denote the square $[\frac{i+3}{7}, \frac{i+4}{7}] \times [\frac{3}{7}, \frac{4}{7}]$ and note that S_0 and S_7 are centred at the same points as the squares $[0, 1]^2$ and $[1, 2] \times [0, 1]$, respectively. Suppose now that the events $E_{0,0}$ and $E_{1,0}$ occur. We then have that no point of X in $\cup_{i=1}^6 S_i$ has more than m points within distance $\frac{3}{7}$. Two points x and y in S_i and S_{i+1} are at distance at most $\sqrt{5}/7$ from each other. Since $\sqrt{5} < 3$, this implies that for all $x \in S_i, y \in S_{i+1}$, there is an edge between x and y. This in turn implies that for all $x \in S_0, y \in S_7$, there is a path from x to y. Similar statements hold whenever two events $E_{l,n}$ and $E_{l+1,n}$ (or $E_{l,n}$ and $E_{l,n+1}$) occur. A simple comparison with independent site percolation on the square lattice now shows that there is an infinite cluster a.s. □

8.4 Stationary spanning forests

Suppose we are given a finite set of points $S = \{x_1, \ldots, x_n\}$ in d-dimensional Euclidean space. A tree $\mathcal{T}$ with vertex set S is called a *spanning tree* for S if each vertex of S is incident to at least one edge of $\mathcal{T}$. A *minimal spanning tree* (MST) for S is a spanning tree such that the sum of the edge lengths is minimal among all spanning trees. If S is such that the distances $|x_i - x_j|$ are all different, then there is a unique MST for S and this tree can be constructed as follows. Start with an arbitrary vertex, x_1 say, and define $T_1(x_1) = \{x_1\}$. Choose the point in S closest to x_1, x_2 say, draw the edge between x_1 and x_2 and define $T_2(x_1) = \{x_1, x_2\}$. Inductively, after having defined $T_k(x_1)$ for some $1 \leq k \leq n - 1$, choose the vertex of $S \backslash T_k(x_1)$ closest to any point in $T_k(x_1)$, draw the edge between these two points and add the new point to $T_k(x_1)$ to obtain $T_{k+1}(x_1)$. This algorithm is called the *greedy algorithm* for obvious reasons and it can be shown that $T_n(x_i)$ is the same for all $1 \leq i \leq n$.

We are now going to describe an infinite and stationary version of this procedure. Note that the notion of a spanning tree (or a spanning *forest*, i.e. a graph with no circuits but not necessarily connected) still makes sense on an infinite set of vertices, but the notion of a minimal spanning tree typically does not. The greedy algorithm itself can still be applied in an infinite set of points. Take a Poisson process X in $\mathbb{R}^d$ with density 1 (the value of the density is unimportant). For any point $x \in X$ we can apply the greedy algorithm described above. This yields, for any integer $n \geq 1$, a tree $T_n(x)$. We write

$$T_\infty(x) = \bigcup_{n=1}^{\infty} T_n(x).$$

Definition 8.2 *The (random) graph $\mathcal{F}$ is the graph with vertex set all points of X and which contains the (undirected) edge $e = (x_i, x_j)$ if and only if e is an edge in either $T_\infty(x_i)$ or $T_\infty(x_j)$.*

It is clear that $\mathcal{F}$ is stationary in the sense that the distribution of the graph structure is invariant under translations. We are interested in the geometry of $\mathcal{F}$ and we shall discuss both local and global properties of $\mathcal{F}$.

Theorem 8.8 *The graph $\mathcal{F}$ is a.s. a forest and all components of $\mathcal{F}$ are unbounded.*

Proof Suppose that $\mathcal{F}$ contains a circuit $(x_1, x_2, \ldots, x_n, x_1)$ with all x_i's different. Suppose that the maximal edge length in this circuit is attained by the edge $e = (x_n, x_1)$, say. Observe that $T_\infty(x_1)$ cannot contain e because the greedy algorithm would first have added all other edges of the circuit. Similarly, $T_\infty(x_n)$ cannot contain e. It follows that $\mathcal{F}$ does not contain e, a contradiction.

Next we show that $\mathcal{F}$ contains only unbounded components. We claim that if $f = (x_1, x_2)$ is an edge of $T_\infty(x)$ for some $x \in X$, then $f \in \mathcal{F}$. This claim is enough since it implies that the component of $\mathcal{F}$ which contains x also contains

$T_\infty(x)$. To prove the claim, suppose that f is an edge of $T_\infty(x)$ and that x_1 is added to $T_\infty(x)$ before x_2 by the greedy algorithm. If f is not an edge of $T_\infty(x_1)$, then $T_\infty(x_1)$ contains only edges which are shorter than $|x_1 - x_2|$. But then, f will never be added to $T_\infty(x)$ which is a contradiction. Hence f is an edge of $T_\infty(x_1)$ and thus also of $\mathcal{F}$. □

The obvious question to be answered here is whether or not $\mathcal{F}$ is a tree. One might guess that a tree is obtained, but this is not so clear, especially if we take into account some results of Pemantle (1991). He shows that 'uniform spanning forests' on the d-dimensional integer lattice can be a.s. trees or forests depending on the dimension.

We end the section with two local properties of $\mathcal{F}$ which could be of some help in understanding the model.

Proposition 8.3 *Suppose without loss of generality that the origin is a point of X and let D be the degree of the origin in $\mathcal{F}$. Then there is a constant c_d, depending on the dimension only, such that $D \leq c_d$. Furthermore, $ED = 2$, in any dimension.*

Proof It is easy to see that in any minimal spanning tree, two edges sharing a vertex cannot make an angle of less then 60 degrees. So for any vertex x there is a uniform bound (in dimension 2 this bound is 6) on the number of edges in $T_\infty(x)$ which have x as an end vertex. Denote these edges by $e_1, \ldots, e_k$, where $e_i = (x, x_i)$. In addition to these edges, x can also be the end vertex of an edge (x, y) in $\mathcal{G}$ for which (x, y) is an edge of $T_\infty(y)$. Denote these edges by $f_1, \ldots, f_n$ where $f_j = (x, y_j)$. We claim that for all $i \neq j$, f_i and f_j make an angle of at least 60 degrees. To see this, suppose not and suppose that $d(x, y_j) > d(x, y_i)$, say. Then $d(y_i, y_j) < d(y_j, x)$ and it follows that also (y_j, y_i) and f_i are edges in $T_\infty(y_j)$. This is a contradiction because $T_\infty(y_j)$ cannot contain a circuit. Finally we claim that f_j and e_i cannot make an angle of less than 60 degrees for any i and j. This follows as in the proof of the first claim after noting that $d(x_i, x) < d(y_i, x)$.

It remains to show that $ED = 2$. The proof is based on a typical volume-boundary argument. Consider the box $B_L = [-L, L]^d$ as usual. Let F_L be the number of components in the graph which we obtain from $\mathcal{G}$ if we only look at points of X in B_L and the connections between them. Let G_L be the number of edges which cross the boundary of B_L, i.e. all edges which have exactly one end point in B_L. (Note that there are a.s. no points of X on the boundary of B_L.) Finally, we denote by $h(x)$ the degree of the vertex

x in $\mathcal{G}$. We now claim that

$$\sum_{x \in X \cap B_L} (h(x) - 2) = G_L - 2F_L. \tag{8.13}$$

This formula seems somewhat mysterious, but it is easy to prove by induction: if X has only one point in B_L the left and right sides of (8.13) are both equal to -2. If we add a point of X in B_L both sides decrease by 2; if we add an edge between two points in B_L then this edge has to be between different components ($\mathcal{G}$ is a forest!) and both sides increase by 2; if we add an edge between a point in B_L and a point outside B_L both sides increase by 1.

We want to take expectations in (8.13). To this end we observe that $EX(B_L) = (2L)^d$ and it is not hard to show with the theory of Palm measures (and intuitively obvious) that $E\left(\sum_{x \in X \cap B_L} h(x)\right) = (2L)^d ED$. Hence we obtain

$$ED - 2 = \frac{E(G_L)}{(2L)^d} - 2\frac{E(F_L)}{(2L)^d}.$$

Using the fact that all components of $\mathcal{G}$ are unbounded we see that $F_L \leq G_L$ and it suffices therefore to show that $E(G_L) = o(L^d)$ for $L \to \infty$. For this, let D_r denote the number of edges at the origin in $\mathcal{G}$ with length at least r. Now G_L counts edges crossing the boundary of B_L and by considering separately those edges with end point in $B_L \backslash B_{L-r}$ and those with end point in B_{L-r} we have, using Proposition 8.3,

$$EG_L \leq c_d \ell(B_L \backslash B_{L-r}) + (2(L - r))^d ED_r$$

whence $\limsup_{L \to \infty} EB_L/(2L)^d \leq ED_r$. Now let $r \to \infty$, and the proof is complete. $\qquad \square$

8.5 Percolation of Poisson sticks

The strength and brittleness of a metal object depends on the fractures present in the material. Typically, the fractures are represented as cracks of varying length and orientation present randomly in the object. In a larger scale, such fractures are also present in geological objects, e.g. fault lines in the study of earthquakes. Although in the first case, the material may be assumed to have homogeneous composition, the geological study will not admit such an assumption of homogeneity. Nonetheless, a simple model used to study such phenomenon is the Boolean model with 'sticks' instead of balls.

Consider a Poisson point process X on $\mathbb{R}^2$ and suppose that each point of the process is the centre of a one-dimensional line (stick) of random length and of random orientation θ with respect to the x-axis. These sticks represent the cracks. Again we assume that the different sticks have an i.i.d. distribution. More precisely, our model consists of points $x_1, x_2, \ldots$ of a Poisson point process with density λ on $\mathbb{R}^2$ and one-dimensional line segments $L_1, L_2, \ldots$ centred at $x_1, x_2, \ldots$ respectively, where L_i has length l_1 and orientation θ_i with respect to the x-axis. We assume that $l_1, l_2, \ldots$ are i.i.d., $\theta_1, \theta_2, \ldots$ are i.i.d. and, for all $1 \leq i, j, l_i$ and θ_j are independent of each other. For this model we may define the critical quantities λ_c, λ_T and λ_S as in the 'ordinary' Boolean model.

Clearly if θ_1 has a degenerate distribution, i.e. all sticks are oriented in the same direction, then no two sticks will intersect a.s. and so percolation will never occur almost surely. Thus for any meaningful study of this model we need to assume that θ has a non-degenerate distribution.

As a simple example let us study the case where

$$\theta_1 = \begin{cases} 0 & \text{with probability } p, \\ 1/(2\pi) & \text{with probability } 1 - p, \end{cases} \tag{8.14}$$

and

$$l_1 = 1 \quad \text{with probability 1.} \tag{8.15}$$

Let L_0 be a stick of unit length centred at the origin and with orientation 0. Let $L_{i_1}, L_{i_2}, \ldots, L_{i_k}$ be sticks which intersect L_0. Clearly, with probability 1, all these sticks are perpendicular to L_0 and are centred in the box $B_{1/2} = [-\frac{1}{2}, \frac{1}{2}] \times [-\frac{1}{2}, \frac{1}{2}]$. Thus k has a Poisson distribution with mean $\lambda(1 - p)$. We call these sticks the first generation sticks. The second generation sticks are all the sticks, except the stick L_0, which intersect the first generation sticks. In a similar fashion we define the $(n + 1)$-th generation sticks as all the sticks, except the sticks which have already been considered in previous generations, which intersect a stick of the n-th generation. Clearly, the expected number of sticks which lie in the component containing L_0 equals the sum of the expected number of sticks at each of the generations. We shall obtain an upper bound of this quantity by placing an independent Poisson process for each stick of the n-th generation and computing the expected number of sticks from this process which intersect the given stick. Adding this expected number over all the sticks of the n-th generation, we obtain an upper bound of the expected number of sticks of the $(n + 1)$-th generation.

Given a stick L of the n-th generation, by placing an independent Poisson process of intensity λ with the orientation and length of sticks given by (8.14) and (8.15) respectively, we have

$$E_\lambda(\text{number of sticks intersecting } L) \leq \begin{cases} \lambda p & \text{if } n \text{ is odd} \\ \lambda(1 - p) & \text{if } n \text{ is even.} \end{cases}$$

Thus, if the n-th generation consists of sticks $L_{j_1}, \ldots, L_{j_m}$, then

E_λ(number of sticks in the $(n + 1)$-th generation$| L_{j_1}, \ldots, L_{j_m}$

are all the sticks of the n-th generation)

$\leq \sum_{i=1}^{m} E_\lambda$(number of sticks intersecting $L_{j_i} | L_{j_i}$ is a stick

of the n-th generation)

$\leq \begin{cases} m\lambda p & \text{if } n \text{ is odd} \\ m\lambda(1 - p) & \text{if } n \text{ is even.} \end{cases}$

Now the number of sticks, m, in the n-th generation is a random variable, and an induction argument assuming that

$$E_\lambda(m) \leq \begin{cases} (\lambda p)^{n/2}(\lambda(1 - p))^{n/2} & \text{if } n \text{ is even} \\ (\lambda p)^{(n-1)/2}(\lambda(1 - p))^{(n+1)/2} & \text{if } n \text{ is odd} \end{cases}$$

$$< \infty$$

yields, on an application of Wald's equation,

E_λ(number of sticks in the $(n + 1)$-th generation)

$$\leq \begin{cases} (\lambda p)^{n/2}(\lambda(1 - p))^{(n/2)+1} & \text{if } n \text{ is even} \\ (\lambda p)^{(n+1)/2}(\lambda(1 - p))^{(n+1)/2} & \text{if } n \text{ is odd.} \end{cases}$$

Hence the expected number of sticks in all generations is at most

$$\sum_{n=0}^{\infty}[(\lambda p)^n(\lambda(1 - p))^n + (\lambda p)^n(\lambda(1 - p))^{n+1}]$$

$$= (1 + \lambda(1 - p)) \sum_{n=0}^{\infty}(\lambda^2 p(1 - p))^n$$

$$< \infty \text{ if } \lambda < \sqrt{\frac{1}{p(1 - p)}}.$$

Thus $\lambda_c \geq \sqrt{1/(p(1 - p))}$. In particular, when $p = \frac{1}{2}$, we have $\lambda_c \geq 2$.

In general, we assume that

$$0 < l_1 \leq R, \tag{8.16}$$

for some $R > 0$, and

$$\theta_1 \text{ has a uniform } [0, 1] \text{ distribution.} \tag{8.17}$$

Under these conditions, not only can we prove the equality of the critical densities λ_c, λ_T and λ_S, we may also define the critical densities via the vacancy structure λ_c^*, λ_T^* and λ_S^* and prove their equality. As in Theorems 3.5 and 4.3 we have

Theorem 8.9   *For the Poisson stick model on the two-dimensional plane, if (8.16) and (8.17) hold then $\lambda_c = \lambda_T = \lambda_S = \lambda_c^* = \lambda_T^* = \lambda_S^*$.*

For higher dimensions, if we consider one-dimensional lines satisfying (8.16) and (8.17) in a higher-dimensional Poisson setting, then it is easy to see that two lines will almost surely never intersect and thus we will not have any percolation. The appropriate analogy will consist of bounded $(d-1)$-dimensional rectangles in a d-dimensional Poisson setting.

8.6 Notes

The results in Section 8.1 are due to Meester and Sarkar (forthcoming) and motivated by a number of papers on discrete fractal percolation (Chayes, Chayes and Durrett, 1988; Chayes and Chayes, 1989; Dekking and Meester, 1990). Theorem 8.3 and Theorem 8.4 are from Molchanov and Stepanov (1983), and Theorem 8.5 is due to Alexander and Molchanor (1994). The equality $p_c + p_c^* = 1$ is from Russo (1978). The material in Section 8.3 is taken from Häggström and Meester (1995), where it is shown that $m_c(d) = 2$ for all d sufficiently large. The basic reference for Section 8.4 is Aldous and Steele (1992). Alexander (1994) has shown, using the occupied version of the RSW theorem, that $\mathcal{F}$ is a tree a.s. in two dimensions. The results of Section 8.5 are taken from Roy (1991).

References

Aizenman, M., and Barsky, D.J. (1987). Sharpness of the phase transition in percolation models, *Commun. Math. Phys.* **108**, 489–526.

Aizenman, M., Kesten, H., and Newman, C.M. (1987). Uniqueness of the infinite cluster and continuity of connectivity functions for short- and long-range percolation, *Commun. Math. Phys.* **111**, 505–532.

Aldous, D., and Steele, J.M. (1992). Asymptotics for Euclidean minimal spanning forests in infinite graphs, *Prob. Th. Rel. Fields* **92**, 247–258.

Alexander, K. (1991). Finite clusters in high density continuous percolation: compression and sphericality, *Prob. Th. Rel. Fields* **97**, 35–63.

Alexander, K. (1994). The RSW theorem for continuum percolation and the CLT for Euclidean minimal spanning trees, preprint.

Alexander, K., and Molchanov, S.A. (1994). Percolation of level sets for two-dimensional random fields with lattice symmetry, preprint.

Athreya, K., and Ney, P. (1972). *Branching processes* (Springer-Verlag, New York).

Berg, J.v.d. (1995). A note on disjoint-occurrence inequalities for marked Poisson point processes, preprint.

Berg, J.v.d., and Kesten, H. (1985). Inequalities with applications to percolation and reliability, *J. Appl. Prob.* **22**, 556–569.

Bezuidenhout, C., and Grimmett, G.R. (1991). Exponential decay for subcritical contact and percolation processes, *Ann. Prob.* **19**, 984–1009.

Breiman, L. (1968). *Probability* (Addison-Wesley, Reading, MA).

Broadbent, S.R., and Hammersley, J.M. (1957). Percolation processes I. Crystals and mazes, *Proc. Cambr. Phil. Soc.* **53**, 629–641.

Burton, R.M., and Keane, M.S. (1989). Density and uniqueness in percolation, *Commun. Math. Phys.* **121**, 501–505.

Burton, R.M., and Keane, M.S. (1991). Topological and metric properties of infinite cluster in stationary two-dimensional site percolation, *Israel J. Math.* **76**, 299–316.

Burton, R.M., and Meester, R. (1993). Long range percolation in stationary point processes, *Random Struct. and Alg.* **4**, 177–190.

Chayes, J.T., and Chayes, L. (1989). The large *N*-limit of the threshold values in Mandelbrot's fractal percolation process, *J. Phys. A: Math. Gen.* **22**, L501–L506.

Chayes, J.T., Chayes, L., and Durrett, R. (1988). Connectivity properties of Mandelbrot's percolation process, *Prob. Th. Rel. Fields* **77**, 307–324.

Daley, D.J., and Vere-Jones, D. (1988). *An introduction to the theory of point processes* (Springer-Verlag, New York).

Dekking, F.M., and Meester, R. (1990). On the structure of Mandelbrot's percolation process and other random Cantor sets, *J. Stat. Phys.* **58**, 335–341.

Denker, M., Grillenberger, C., and Sigmund, K. (1976). *Ergodic theory on compact spaces* (Springer-Verlag, New York).

Dunford, N., and Schwartz, J.T. (1958). *Linear Operators* (Wiley–Interscience, New York).

Feller, W. (1978). *An introduction to probability theory and its applications* (Wiley, New York).

Fortuin, C., Kasteleyn, P.W., and Ginibre, J. (1971). Correlation inequalities on some partially ordered sets, *Commun. Math. Phys.* **22**, 89–103.

Gilbert, E.N. (1961). Random plane networks, *J. Soc. Indust. Appl.* **9**, 533–543.

Grimmett, G.R. (1989). *Percolation* (Springer-Verlag, New York).

Grimmett, G.R., and Marstrand, J.M. (1990). The supercritical phase of percolation is well behaved, *Proc. R. Soc. London* **430**, 439–457.

Grimmett, G.R., and Stirzaker, D.R. (1992). *Probability and random processes* (Oxford University Press, London).

Häggström, O., and Meester, R. (1995). Nearest neighbour and hard sphere models in continuum percolation, preprint.

Hall, P. (1985). On continuum percolation, *Ann. Prob.* **13**, 1250–1266.

Hall, P. (1988). *Introduction to the theory of coverage processes* (Wiley, New York).

Harris, T.E. (1960). A lower bound for the critical probability in a certain percolation process, *Proc. Cambr. Phil. Soc.* **56**, 13–20.

Kertesz, J., and Vicsek, T. (1982). Monte Carlo renormalization group study of the percolation problem of discs with a distribution of radii, *Z. Physik B* **45**, 345–350.

Kesten, H. (1980). The critical probability of bond percolation on the square lattice equals $\frac{1}{2}$, *Commun. Math. Phys.* **74**, 41–59.

Kesten, H. (1982). *Percolation theory for mathematicians* (Birkhäuser).

Krengel, U. (1985). *Ergodic theorems* (DeGruyter, New York).

Meester, R. (1994). Uniqueness in percolation theory, *Stat. Neerl.* **48**, 237–252.

Meester, R. (1995). Equality of critical densities in continuum percolation, *Adv. Appl. Prob.*, to appear.

Meester, R., and Roy, R. (1994). Uniqueness of unbounded occupied and vacant components in Boolean models, *Ann. Appl. Prob.* **4**, 933–951.

Meester, R., Roy, R., and Sarkar, A. (1994). Non-universality and continuity of the critical covered volume fraction, *J. Stat. Phys.* **75**, 123–134.

Meester, R., and Sarkar, A. (forthcoming). On continuum fractal percolation, in preparation.

Menshikov, M.V. (1986). Coincidence of critical points in percolation problems, *Soviet Mathematics Doklady* **33**, 856–859.

Menshikov, M.V., and Sidorenko, A.F. (1987). The coincidence of critical points in Poisson percolation models, *Th. Prob. Appl.* **32**, 547–550.

Mode, C.J. (1971). *Multitype branching processes: theory and applications* (American Elsevier Publishing Company, New York).

Molchanov, S.A., and Stepanov, A.K. (1983). Percolation in random fields I,II, II, *Theor. Math. Phys.* **55**, 478–484, 592–599 and **67**, 434–439.

Pemantle, R. (1991). Choosing a spanning tree for the integer lattice uniformly, *Ann. Prob.* **19**, 1559–1574.

Penrose, M.D. (1991). On a continuum percolation model, *Adv. Appl. Prob.* **23**, 536–556.

Penrose, M.D. (1993). On the spread-out limit for bond and continuum percolation, *Ann. Appl. Prob.* **3**, 253–276.

Penrose, M.D. (1995a). Single linkage clustering and continuum percolation, *J. Multivariate Anal.* **52**, to appear.

Penrose, M.D. (1995b). Euclidean minimal spanning trees and continuum percolation in high dimensions, preprint.

Penrose, M.D. (1995c). Continuity of critical density in a Boolean model, preprint.

Penrose, M.D., and Pisztora, A. (1994). Large deviations for discrete and continuous percolation, preprint.

Petersen, K. (1983). *Ergodic theory* (Cambridge University Press, Cambridge).

Phani, M.K., and Dhar, D. (1984). Continuum percolation with discs having a distribution of radii, *J. Phys. A: Math. Gen.* **17**, L645–L649.

Pugh, C., and Shub, M. (1971). Ergodic elements of ergodic actions, *Comp. Math.* **23**, 115–122

Reimer, D. (1994) Butterflies, preprint.

Roy, R. (1988). The Russo-Seymour-Welsh theorem and the equality of critical densities for continuum percolation on $I\!R^2$, Ph.D. Thesis, Cornell University.

Roy, R. (1990). The RSW theorem and the equality of critical densities and the 'dual' critical densities for continuum percolation on $I\!R^2$, *Ann. Prob.* **18**, 1563–1575.

Roy, R. (1991). Percolation of Poisson sticks on the plane, *Prob. Th. Rel. Fields* **89**, 503–517.

Roy, R., and Sarkar, A. (1992). On some questions of Hartigan in cluster analysis: an application of BK-inequality for continuum percolation, preprint.

Rudin, W. (1970). *Real and complex analysis* (McGraw-Hill, New York).

Rudolph, D. (1979). Smooth orbit equivalence of ergodic $I\!R^d$-actions, $d \geq 2$, *Trans. Am. Math. Soc.* **253**, 291–302.

Russo, L. (1978). A note on percolation, *Z. Wahrsch. verw. Geb.* **43**, 39–48.

Sarkar, A. (1994). Some problems of continuum percolation, Ph.D. Thesis, I.S.I., New Delhi.

Sarkar, A. (1995). Continuity and convergence of the percolation function in continuum percolation, preprint.

Seymour, P.D., and Welsh, D.J.A. (1978). Percolation probabilities on the square lattice, *Ann. Discr. Math.* **3**, 227–245.

Tanemura, H. (1993). Behavior of the supercritical phase of a continuum percolation model on $I\!R^d$, *J. Appl. Prob.* **30**, 382–396.

Whitney, H. (1933). Planar graphs, *Fund. Math.* **21**, 73–84.

Zuev, S.A., and Sidorenko, A.F. (1985). Continuous models of percolation theory I, II, *Th. Math. Phys.* **62**, 51–58, 171–177.

Index

action, 23, 24

BK inequality
  continuum, 35
  discrete, 7, 165
bond, 3
Boolean model
  dependent, 221
  general, 15, 184
  Poisson, 7, 40, 92, 122
bounds on λ_c, 52, 89
branching random walk, 89

circuit, 3
closure, 187
cluster, 3
  size, 89
complete coverage, 41, 184
component
  occupied, 15
  vacant, 15
compression, 127
connection function, 18, 155
continuity
  of critical density, 71
  of percolation function 77, 78, 119
coupling, 28
covered volume fraction, 122, 209
  at criticality, 123
critical densities, 45, 50, 53, 92, 152
  equality of, 53, 59, 109, 159
critical probability, 3
crossing probability
  continuum, 33
  discrete, 4
cutting and stacking, 199
  generalised, 202

density, 11, 25
  relative, 127
differential inequality, 163
discrete percolation, 2
discretisation, 2, 32, 52, 59, 68, 161, 175,
  220, 225
disjoint occurrence, 6, 34
dual graph, 5

event
  decreasing, 6, 31
  increasing, 6, 31
ergodic decomposition, 183
ergodic theorem, 22, 23, 24
ergodicity, 25
  of Boolean model, 27
  of Poisson process, 26
  of random-connection model, 27
exponential decay, 68
exterior, 187

fidi distribution, 10
finite correlation, 218
FKG inequality
  continuum, 32
  discrete, 6
forest, 223, 226
fractal, 209

Gaussian field, 219
greedy algorithm, 227

high density, 127, 140, 173
high dimensions, 88, 89
Hilbert–Schmidt operator, 86

infinite volume limit, 168
interior, 187
invariant, 21

Kuratowski's criterion, 190

lattice, 3
Lebesgue density theorem, 153, 201
Lebesgue set, 153
level set, 217

martingale convergence, 32
measure-preserving (m.p.)
    dynamical system, 21
    transformation, 21
mixing, 26

neighbour, 3

occupied region, 15

Palm distribution, 13
path, 3
percolation
    bond/site, 8
    discrete, 2
    fractal, 209
    level set, 216
    in one dimension, 43
percolation function, 3, 18
    continuity of, 77, 78, 119
    monotonicity of, 3, 29
phase transition, 2, 4, 53
pivotal, 7
planar graph, 190
point process, 9
    stationary, 10
Poisson Boolean model, 17
Poisson process
    homogeneous, 11
    non-homogeneous, 12

random-connection model
    dependent, 225
    general, 18, 197
    Poisson, 151
random field, 216
rarefaction, 139
relative density, 127
RSW lemma
    occupied, 121
    vacant, 96
Russo's formula, 7, 8, 163, 167

scaling, 29
spanning forest, 226
spanning tree, 226
stationary, 10
sticks, 229
Stirling's formula, 128
subcritical, 18
supercritical, 18
superposition, 11

tangent balls, 224
thinning, 13
T-invariant, 21
tree, 204, 206, 226
triangular lattice, 87

uniqueness, 5
    in general Boolean model, 194
    in general random-connection model, 197
    in Poisson Boolean model, 63, 116
    in Poisson random-connection model, 172

vacant region, 15

weak convergence, 71